PLAYING THE PONIES and OTHER MEDICAL MYSTERIES SOLVED

PLAYING THE PONIES and OTHER MEDICAL MYSTERIES SOLVED

Stuart B. Mushlin,
MD, FACP

RUTGERS UNIVERSITY PRESS MEDICINE
New Brunswick, Camden, and Newark, New Jersey, and London

Library of Congress Cataloging-in-Publication Data

Names: Mushlin, Stuart B., author.

Title: Playing the ponies and other medical mysteries solved/
   Stuart B. Mushlin, MD, FACP

Description: New Brunswick, New Jersey: Rutgers University Press, [2017]

Identifiers: LCCN 2016025798| ISBN 9780813570563 (hardcover : alk.
   paper) | ISBN 9780813570556 (pbk. : alk. paper) | ISBN 9780813570570
   (e-book (web pdf)) | ISBN 9780813575155 (e-book(epub))

Subjects: LCSH: Medicine-Practice. |Physician and patient. | Diagno-
   sis.| Medicine –Case studies.

Classification: LCC R728.M875 2017 | DDC 610.68—dc23

LC record available at https://lccn.loc.gov/2016025798

A British Cataloging-in-Publication record for this book is available
   from the British Library.

∞ The paper used in this publication meets the requirements of the
American National Standard for Information Sciences—Permanence
of Paper for Printed Library Materials, ANSI Z39.48–1992.

www.rutgersuniversitypress.org

Manufactured in the United States of America

FOR FRANCINE,
ALWAYS

CONTENTS

PREFACE

LIKE SO MANY YOUNG PEOPLE, I WAS TOTALLY ignorant of the reality of being a physician. I entered medical school in 1969 with the vague notion that I wanted to help people and an even vaguer concept that I had the capability to learn to do the job. There were supporting factors, which I would have acknowledged if asked. Both of my parents had suffered from tuberculosis in the pre-antibiotic era, which affected them deeply. They both recovered, with my father devoting his career to the New York Lung Association and my mother working as a nurse. My parents had great respect for the doctors who helped cure them, and on some level I wanted to honor them by becoming a physician. And, truthfully, medicine was also a predictable and "safe" career path with little uncertainly about how life would be organized for a large number of years.

But I totally missed the concept that medicine is truly a calling: A committed physician makes a covenant with his patients and his colleagues to strive, daily, to be the best he or she can be. This is a wonderful fact, but a heavy burden. To balance the demands of essential continuous learning, patient care, and family life is no easy task. Many doctors never find that balance. Indeed, when I matriculated, the concept of balance in one's life as a physician was not even discussed. The vast majority of our role models lived and breathed medicine twenty-four hours a day.

So much of medical education and practice has evolved for the better. When I was a student my medical school could find only four acceptable women a year (in a class of ninety) for enrollment. Needless to say, these women were head and shoulders more qualified than the average male classmate, but there was a belief that, because they might have children or leave the workforce, they would not use their medical degrees as well as their male counterparts. And I suspect some faculty felt that women were too weak for the rigors of a life as a physician. Fortunately, that foolishness is long gone, and women now constitute over half of the students in medical school. Many applicants now have real-life experience prior to starting medical school; they bring a maturity and broader worldview to undergraduate medical education than was present in my generation.

Medical school admissions have become increasingly selective even though there are now many more medical schools in the United States than there were forty years ago. It is very difficult to get accepted unless one's grades are truly outstanding. And many applicants have extensive portfolios that include components that seem to have left little time for anything but work and community service. Such students are likely to be highly organized in medical school and able to manage their time well, but they are also probably going to be more concerned about grades and the opinions of others than about being self-actualized and reflective about medicine, medical training, and the human condition.

I regard myself as fortunate that criteria were far less stringent when I was applying to medical school. My grades were good but not great. I took a lot of courses outside of the prescribed premedical curriculum. And I took courses in many disciplines that simply interested me. Intellectually experimenting, as I did as an undergraduate, sometimes meant that the course and I didn't mesh well—and my grade reflected it. I eventually majored in English as I enjoyed reading and writing and felt that novels, plays, and poetry provided real insights into universal human concerns. From my current vantage point, my undergraduate education was superb preparation for the practice of medicine.

Though a lot of math and chemistry is still required for admission to medical school, as a practitioner I use very little of it in my daily life. But I certainly, moment by moment, learned to listen to patients and apprehend the theme of their visit. Were they sad, were they miserable, were they worn down by their illness? Were they resilient, were they resigned, were they content? The practice of medicine offers a parade of individuals passing through a doctor's life, each one presenting a story. Stories about their children, their jobs, their social standing, their fears, their embarrassments, their indiscretions. And, as I hope this book illustrates, these patients bring vivid impressions to the doctor. Through their stories, my patients reminded me daily of the human condition, the roles of luck and resilience in life, and, occasionally, the nobility of those who must confront painful illness.

I have personalized my patients' stories but disguised them to protect their privacy, and I call them by false initials. All the stories are fundamentally true, and I was the primary care doctor for each of them. I have made no effort to cover certain diseases or certain character types comprehensively. Rather, I've tried to share some of the vivid memories of the people I've been privileged to try to help.

I have made mistakes in my years of practice, something that I am not proud of. But I haven't tried to hide my missteps or confusion in these chapters. Doctors deal with uncertainty all day long, and this fact is underappreciated by the public. Furthermore, we are human, and on some days, we're not at our best. What I hope to convey in this book is a realistic and genuine portrayal of the pleasure and joy of being a privileged witness to the human condition.

Brookline, Massachusetts

April 2016

PLAYING
THE PONIES
and OTHER
MEDICAL
MYSTERIES
SOLVED

BACK PAIN

WHEN I FIRST MET MR. M., HE WAS AN IMPOSING figure, and it was clear he was trying to intimidate me. He was over six feet tall and heavyset, with very fair, freckled skin and a thin patch of orange hair. He said, "So you're my new doctor, eh?"

Accompanying him was his wife of forty years. She was a diminutive and quiet woman with obvious rheumatoid arthritis affecting her hands and legs. It was clear she let him be the boss.

They were referred to me by her rheumatologist, who was an old colleague and close personal friend. Like many patients in our large hospital, they wanted a doctor on-site to coordinate their many specialists and because they had a certain level of comfort with the institution.

Mr. M. had been raised in a blue-collar, mostly Irish American, part of Boston. Like his father before him, he had become a Boston policeman. But he was unlike his father in more ways than one: He had risen in the ranks, and in a branch of the Boston Police Department loathed by the average patrolman—he had been, for many years, in charge of Internal Affairs. As the Thin Blue Line took great pride in covering the backs of fellow officers, Internal Affairs was loathed. It was clear my new patient was one tough customer.

When he saw me, he needed an internist after a recent hospitalization for a serious cardiac problem. He had retired from the Boston Police Department and was working as a security consultant. His wife and he had long planned a fortieth-anniversary trip to Paris. His wife's rheumatoid arthritis, which significantly limited her mobility and stamina, required special considerations for the trip. They had carefully planned their hotels and itinerary. It was to be the trip of a lifetime for this essentially blue-collar couple who, I would learn, had simple tastes and strong moral values.

So it was a surprise when, while walking on the Champs-Élysées, it was Mr. M. who collapsed on the sidewalk. He had had a heart attack. He received excellent attention from the Parisian first responders, and a thoroughly modern French hospital, which provided a few days of excellent care, served as the last hotel on their trip.

Arriving back in the United States, they sought care from one of my colleagues in cardiology. He determined that, not only had Mr. M. had a heart attack, he also had a weakened heart, a condition called cardiomyopathy. It wasn't clear if the cardiomyopathy was caused by blockage of his heart arteries or whether it was from other possible causes (of which there are many). Treatment is fairly standard; the usual medications are started, and time will tell how the patient will do. In Mr. M.'s case, one of the medical interventions was placement of a pacemaker and an implanted defibrillator. Mr. M.'s cardiologist thought he should have an internist, and Mrs. M.'s rheumatologist referred him to me.

It soon became clear that Mr. M.'s imposing and authoritarian presence was a cover for the fact that he was frightened of, and frustrated by, his sudden illness. He had enjoyed wonderful health for many years—he took no medication and was quite prudent in his habits—and he was shocked and dismayed that their wonderful vacation had ended this way. As we talked, he relaxed more. He often included his wife in his responses, very sweetly calling her "Mother." He was forthcoming but made it clear that he was to be respected. In addition to his wife, his other point of pride was their son. A product of the Boston school system, his son was a scholar and athlete who had gone to the prestigious Boston Latin School and then Harvard College, where he had played two varsity sports and graduated with distinction. Seeing him speak so gently to his wife and defer to her answers to some of my questions, and noting the intense pride with which he spoke about his son, quickly gave me a lot of insight into his character.

We developed a rapport and, for four or five years, I was quite worried about his cardiomyopathy, as I had a number of patients with that diagnosis whose conditions had relentlessly progressed to severe and, finally, fatal heart failure. But Mr. M. was in a different category; his cardiomyopathy improved significantly over the years. The heart murmur he had as a result of his enlarged and dilated left ventricle decreased over time. His defibrillator only fired twice, and those were both in the first year of its implantation. I got to know Mr. M. and his wife quite well, though she never became a patient of mine. I met their devoted son, who was indeed as wonderful as he was billed. I got a real sense of their lives as a family. In fact, they were living examples of how the children and grandchildren of immigrants can flourish in our country, if given opportunity and education. This family represented all that was fine in the middle class of our great city.

Cardiovascularly, Mr. M. was doing fine. His biannual checkups were always stable. But he started gaining weight and developed adult-onset diabetes. With Mother in the room as a witness, I explained to him the consequences of diabetes if left untreated and how—when diet modification and prudent exercise were used together to cause weight loss—all of these risks could be mitigated, or even reduced to zero.

He got the message. In six weeks, he had lost ten pounds, and in another eight weeks, he had lost another ten pounds. I was delighted, and so, too, was his pancreas, as his glucose levels were now normal. I cheered both him and Mother for their wonderful work and planned to see him again in three months, convinced that his weight loss was real and likely durable.

Unusually, he called for an appointment just a month after his last visit. He had back pain. He had never experienced back pain, even in his years as a cop on the beat, although admittedly he was a lot younger then.

He described the pain as intense. He reported it was in the low back. It radiated to the left buttock much more than the right buttock, but it also went to the high right buttock. He stated it was bad all day and probably worst when he first got out of bed. His examination was not revealing: He had normal tone in his proximal (near) and distal (far) legs, no sign of muscle atrophy (thinning and weakening), and reflexes were equal and slightly diminished (he scored 1+ on the reflex scale; scores of 1 to 3+ are in the normal range). His sensory examination showed no loss of peripheral nerve sensation, but he did have some mild loss of vibratory sense in both feet. He could squat to the floor and, with just a little help, get up on his own, but squatting was so painful that this proud and stoic man would yelp when he did it.

The examination just described is used to look for signs of nerve irritation or damage. Peripheral nerves exit the spinal cord and course outward to the organs and eventually the skin. When nerves are pinched or damaged, the muscles they supply lose bulk and become weaker. The sensation that nerves normally transmit is diminished when the nerves are compressed. And last, the automatic responses that are the reflexes are diminished or lost when a peripheral nerve is damaged. The exam didn't pinpoint a specific nerve as a problem and only implied that he had some symmetric, age-related loss of sensation, something quite common for a patient over the age of seventy.

Back pain is the second most common complaint (after elevated blood pressure) in a general practitioner's office. I have seen many cases of it. Because of his age and its rapid onset, I assumed this was from canal stenosis (narrowing) of the lower spine. I recommended a nonsteroidal anti-inflammatory drug for a ten-day course. Should he not improve, I would reexamine him, looking for changes, and request X-rays of the lower spine. Surprisingly, he called three days later, in much more pain. I prescribed a very small amount of a low-dose codeine/acetaminophen drug and asked him to stay in touch. He called again in two days, stating the pain was increasingly intense and the low-dose narcotic wasn't helping at all.

On reexamination in the office, he clearly was in significant pain, both while walking back with me to the examining room and while seated on the examination table. He was nearly crying with pain—a startling thing to see in this man I knew so well. Reexamination was unchanged but for his degree of pain. I immediately ordered plain X-ray films of his back, but they showed only osteoarthritis of the lower spine and no sign of a fracture or malignancy. Thinking that pain this severe could conceivably be from a slowly leaking lower abdominal aorta, I also ordered an MRI scan of his abdomen and blood vessels, with special focus on the aorta. Although his aorta wasn't that of a twelve-year-old, and had some calcification and irregularity, there was no sign of an aneurysm or leak.

I was puzzled and concerned. So I asked him about the history of this illness again. It started one morning when he awoke. He had excruciating pain merely getting out of bed in his lower back and glutei (buttock muscles), on the left much more so than the right. The pain was present all day but definitely worse after sleeping. When I asked him to describe the pain, it suddenly became clear to me that he was feeling stiff and sore (like that after excessive physical exertion) rather than having sharp or piercing pain. There was no numbness, tingling, or heat sensation—all symptoms that suggest nerve irritation. There was no pain over the bones of the spine. When I asked him directly if he had the same sensation in the shoulder, he replied that he did, but much less so. When I asked him how long the shoulders had been involved, he replied he thought maybe only in the last week.

It was now clear to me that this process was not localized to the lower back and that the pain was actually severe stiffness, not the typical pain symptoms I was used to seeing with run-of-the-mill lower back pain. I became suspicious that this was a presentation of polymyalgia rheumatica. I sent off for a high-sensitivity c-reactive protein (hs-CRP) lab test. It came back with a value of 96, very elevated, and consistent with that diagnosis.

Polymyalgia rheumatica is a disease of unknown cause that primarily affects people over the age of fifty. In my clinical experience, it rarely affects people under the age of seventy. It is characterized by morning stiffness in the pelvic girdle and shoulder girdle areas. One or both of these areas may be affected. If both are affected, they can be affected simultaneously, or one can be followed and cumulated by the other area. No one knows the cause. It could be a manifestation of a more malicious condition, temporal arteritis, in which the vessels of the scalp and cranium are inflamed. Temporal arteritis needs to be carefully excluded when a patient presents with the symptoms of polymyalgia.

Pure polymyalgia, not as a manifestation of another condition, tends to be self-limited, but if untreated it usually lasts for around a year and a half. And untreated patients are miserable. They experience pain, often lose mobility, and frequently lose weight. In fact, untreated patients with polymyalgia rheumatica can look like they have a terminal malignancy. The disease is treated with prednisone in moderate doses, and the response to prednisone is dramatic. If the patient doesn't feel much better within forty-eight hours on 15–30 milligrams (mg) of prednisone (a moderate dose), the diagnosis is really in doubt. Prednisone levels are tapered down slowly over six to nine months unless symptoms recur, at which time the lowest dose that controls the symptoms is used.

Although I have seen and treated many patients with this condition, Mr. M.'s case illustrates how it can fool the experienced internist. I thought he had typical low back pain from osteoarthritis or spinal stenosis. His lack of the descriptor "stiff" let me go down the wrong path. As his symptoms worsened, I was so taken aback by his unrelenting and progressive pain, literally making this grown man cry, that I neglected to go back and take his history again. When I did and let him freely associate words to describe what he was feeling, the diagnosis suddenly became clear. Most people with the illness say, "I feel stiff" or "If I sit in one spot for thirty minutes I feel very stiff when I have to get up." The most dramatic description of this stiffness came from a man's wife, who told me she had to hoist her husband out of bed every morning by yanking on his arms while her husband was screaming in pain.

I treated Mr. M. with 20 mg of prednisone initially, and I spoke with him the next day. He felt nearly normal by noon the next day. This let me know we were dealing with the correct diagnosis. Over the next five months, his prednisone was slowly tapered to 5 mg every morning with control of his symptoms. Because of his self-discipline, he managed to avoid becoming diabetic again while on the higher doses of prednisone. This truly impressed me, as weight gain is a side effect of prednisone.

When I last saw him, he was feeling quite well. However, I noticed that his long-standing aortic cardiac murmur was now harsher and louder, and I was concerned that he was developing a significant narrowing of the aortic valve area. I sent him for an echocardiogram, but that is a story for another time.

IT'S A SMALL WORLD

A.S. WAS A FORTY-FOUR-YEAR-OLD INFORMATION technology (IT) worker at our hospital who had been in good health all of his life. Born in the Philippines, he immigrated to the United States at the age of ten. He grew up in Houston and earned a degree from Rice University. He had been employed by our hospital IT group for the last eight years. When your computer froze or you needed help with a network program, he was often the one who showed up and fixed it. Any time I had needed his help, he always successfully solved the issue.

So I felt terrible that I could not diagnose, or successfully treat, his medical problem. I was not his primary care physician, but he was referred to me for chronic, persistent knee arthritis.

He was single and had lived locally for the last fifteen years. He did not smoke and only occasionally drank, and never to excess. He was bisexual and when he had sexual relations with either gender, he was careful, and, to his knowledge, he had never had a sexually transmitted disease. Both of his parents had immigrated with him, and both were now deceased. His mother, who had been a nurse, had diabetes and kidney failure. His father, who had been a manual laborer and a smoker all his life, died after a second heart attack. A.S. had two younger sisters, both of whom were in good health. He did not exercise, but his weight had been stable for over ten years, and he was quite slender. He denied bowel problems such as colitis. He had no back pain. He had no rashes or psoriasis that he was aware of. He had no cough, muscle pain, Raynaud's phenomenon (a condition in which the digits overreact to the cold and turn color), or fever. He had no cardiac symptoms. No other joints were swollen. Because he worked in the hospital, he had a skin test for tuberculosis yearly, and it was always negative. He had no recollection of a tick bite or the classic target-type lesion of Lyme disease.

The joint pain and swelling were quite insidious. He had not been exercising or working on his knees. One morning, he noticed the knee was stiff. He thought little of it and went to work. Over the next month, the knee continued to be stiff in the morning hours, but after completing his morning bathing and shaving routine, he was no longer troubled by it, and he wouldn't think about it at all until the next morning, when the stiffness was notable again. He had no pain. At most, he felt a sense of tightness within the knee, but he took pains to clearly state that it wasn't painful.

Over time, he became acclimated to this morning discomfort and stiffness, and he didn't think about it much. He did try some over-the-counter nonsteroidal anti-inflammatory medications, but they didn't help, so he discontinued them. The knee issue had been going on for over a year when he noticed that the knee was now, subtly, also swollen. Again, it was not painful or tender, merely stiff. But because of the swelling, he could not fully bend his knee. Though none of these signs and symptoms were problematic for him, he had a routine exam with his primary care doctor, who suggested he see me.

The rheumatologist's evaluation of the isolated and chronic swollen knee is a common situation. Like many common presentations in the field of medicine, we have a checklist to go through. Often, the history and physical examination are not helpful for a diagnosis, and then it becomes more art than science as to how to treat the patient.

Common causes for one large arthritic joint are Lyme disease (if the patient is in the appropriate geographic region), arthritis caused by inflammatory bowel diseases, and infectious causes—and in a very slowly accumulating single large joint, we think of more indolent (slow to develop) infections. Last, it can be caused by joint space tumors, either of bone or the lining of the joint (the joint lining is the synovium, and inflammation of the synovium is called synovitis). We commonly find disruption to the joint architecture from old trauma (an orthopedic issue).

The patient's exam was remarkable only for his swollen left knee. He was a slender man in no distress. He was afebrile (he had no fever), and he had a chronically low blood pressure, which caused him no problems. There was a small red patch on the high thoracic spine region of his back, which had been there for years but didn't bother him. His lungs were clear, his cardiac examination was entirely normal, and his neurologic examination was normal. There was no enlargement of his liver, his spleen was not palpable—and therefore not likely enlarged—and he had no swelling of any lymph nodes. Psychologically, he was relaxed about his condition, and in general, he was relaxed about his health; he certainly wasn't a worrier.

I initiated the usual workup and evaluation. X-rays showed soft-tissue swelling and no joint destruction or classic lesions. Lab work showed no anemia and a normal white blood cell count, with a normal distribution of white cells among the various cell lines. Specific rheumatologic testing was negative for the rheumatoid arthritis factor (anti-CCP antibody), negative for the lupus screen test (the antinuclear antibody, or ANA), and negative for elevation of uric acid (a necessary but not sufficient clue to the diagnosis of chronic gout). A measure of inflammation, hs-CRP, was also normal. Lyme Western blot testing (a test for Lyme disease) was negative.

With a benign knee X-ray and unrevealing lab tests, I aspirated (removed fluid from) his knee joint on the next visit. The fluid was slightly straw yellow in color and had a slight decrease in normal viscosity. There were 1,500 white blood cells, and most were mononuclear cells. The protein level in the fluid was slightly raised. A Lyme disease polymerase chain reaction (PCR) test to check for sequestered Lyme disease in the knee was negative.

All of the preceding evaluations yielded very nonspecific results. There was evidence of mild inflammation in the joint but no sign of a classic rheumatic disease and no sign of systemic inflammation. The mildly inflamed, chronically swollen knee is a common scenario that rheumatologists encounter. I chose to manage it in a way that I knew has been moderately successful. Other doctors might approach therapy differently; there is no one definitive best way to approach this common arthritic condition. I offered A.S. a mild antirheumatic agent called hydroxychloroquine. It is derived from an old antimalarial drug and has weak activity against various forms of arthritis. Some patients respond very well; others, not at all. The drug takes a good eight weeks to provide benefit, so the patient must be counseled to wait—which for this patient was not a problem. There are side effects to all medications, and this drug has a few serious ones, so I spent plenty of time discussing what the side effects were and how to monitor them. He took the medication and came back in just under three months' time, during which he'd noticed no improvement.

It is my habit that, when hydroxychloroquine alone doesn't work, I then add on medications, administering the least toxic medicines I know. In this case, I added a sulfa drug called sulfasalazine that, when it works, also requires a few months to make a difference. In this circumstance, I now had to inform him of an additional panoply of potential side effects.

Two months later, there was still absolutely no progress. I reexamined him at that time, which is always a good idea when a patient shows no improvement. Sometimes an illness will have progressed, and we can then make new diagnostic conclusions with the new findings. At other times, there will have been no apparent change. A lack of progression after a few months' time, although not ideal, is a modest sign of possible stabilization.

On reexamination, A.S. had a cool, mildly swollen knee joint. There was no change in the range of motion of the knee, and the knee was not painful. There were no skin changes over the knee, and the only positive skin finding was the persistent red patch on his back. We examine the skin mainly to look for subtle signs of psoriasis because psoriasis is occasionally associated with inflammatory arthritis.

As I was at a loss, I suggested to A.S. that I would like a colleague with expertise in dermatology and rheumatology to see him. Getting a second opinion when a patient is unimproved is always a good idea. I wasn't optimistic that the second opinion would offer anything, but I greatly respected my colleague's opinion.

A.S. agreed. In truth, I was probably more bothered by his problem than he was, but he was willing to get a second opinion. I asked him to keep taking his current two medications. We agreed that if there was no improvement, he would return in a month or so, and we would fully discuss the addition of more toxic medication options and their risks and benefits.

I called my dermatology colleague and frankly asked her to biopsy the rash, even if she thought the it was nothing, because it was the only other positive physical finding for my patient and the only finding that hadn't been explored. I almost never tell a colleague what to do during a consultation, but, fortunately, she didn't take offense and did the requested biopsy.

The biopsy came back teeming with a type of bacteria called Hansen's bacillus, the causative agent of leprosy! I had never seen a case of leprosy in all my years of practice, so leprosy as a cause of chronic arthritis certainly wasn't in my mental data bank. So I did what doctors do and went to the literature. Sure enough, leprosy can cause a couple of types of arthritic conditions, and my patient's findings and clinical course were consistent with a diagnosis of leprosy arthritis.

I told A.S. the diagnosis and encouraged him to see a regional infectious disease expert with a strong interest in leprosy, who put him on a two-drug antileprosy regimen. Unfortunately, A.S. had a very aggressive response to the treatment, and he developed fevers and painful arthritis in many joints. The infectious disease expert knew how to manage this unfortunate response to treatment. It's now six years later, and the patient's knee arthritis is no longer present. But it took over two years for it to go away, and there is the distinct possibility that he will have premature osteoarthritis of that knee because of the low-grade and chronic inflammation present in the knee for a number of years. Another outcome, and a less happy one, is that the treatment regimen, in addition to the arthritis flare and fevers, caused his skin to darken, particularly on his face. This has not improved in the remaining six years.

The case is illustrative of a number of things. We have a saying in medicine that, "when you hear hoof beats, think horses and not zebras," which means that common things occur much more often than uncommon things. A.S. had a zebra, and it was so arcane I had never even heard of it as a cause of arthritis. Another point is that he genuinely suffered from the treatment; his face became quite dark and in the early months of therapy, he endured a systemic reaction to the treatment and felt quite unwell. Though he never said so, I strongly suspect he blamed me, and my well-intended interest in his condition, for all these side effects. In truth, he probably would not have minded living with the original condition. As to the rash, had I examined it more closely, I might have made the diagnosis of leprosy much earlier. A characteristic finding of the flat, red rash seen in leprosy is that the area of the rash lacks sensation—had I touched the rash with a pin or the end of a Q-tip, I might have realized this was not a run-of-the-mill rash. Lastly, it's not entirely clear how he developed leprosy. It's a slow-growing illness, and usually requires repeated and prolonged close contact. He did grow up in a crowded home in the Philippines and shared a bed with an aunt and uncle for a number or years, but, to his knowledge, none of his family has or had leprosy.

Ultimately, A.S. was cured of his problem, but was it worth it? He voted clearly by never returning to my office. For me, it was a lesson that dogged persistence by the doctor may make the doctor feel like he is helping the patient, but great patient care is not exclusively about getting the right answer.

EVERYTHING REALLY CAN GO WRONG IN THE HOSPITAL

HE WAS A SPOOK. IT WAS NEVER CLEAR TO ME what branch of government W.S. was in; all he would reveal is that he knew a lot about nuclear weapons and had worked with the State Department developing concepts for global and regional nuclear disarmament. He traveled by military air transport when he flew, so he must have been a somebody.

He was in his late fifties and a Yale Man—yes, some people of a certain stripe and vintage still would unabashedly use that term. Chronically slightly disheveled and moderately overweight, he usually appeared in a wrinkled J. Press blue blazer, gray slacks, and brown cap-toe shoes.

He was married to a highly intelligent Brit, who, for reasons never clear to me, was raised in Austria. She was considerably younger, and their relationship was testy—they would disagree and argue over the smallest points when they were in my office together. But it was clear that, in spite of the verbal sparring, they shared deep and abiding love and respect.

He found me by way of an old teacher of mine who was his physician in New York City. To me, this is rarely a good thing. Rather than being flattered that I have been recommended, I usually find that I can never measure up to the advanced billing or that the patient is really a New Yorker at heart and finds Boston simply too parochial. The bonds that unite the doctor and patient tend to be made of silk and not jute in these referral relationships.

He was not a healthy guy. He had had two prior cardiac bypass surgeries, the first in Washington, D.C., back when many people who could have managed their condition medically had had cardiac bypass surgery instead. The second bypass was in the late 1990s and was essentially a redo of his prior surgery after he had sustained a small heart attack. While the surgeons were fixing his coronaries, they also gave him a new aortic valve to replace his worn original valve. Not surprisingly, this very sharp fellow felt that he had lost some brain cells after the second surgery. I investigated his New York City records and found that, indeed, he had been on the bypass pump for a few hours, and often these patients evince some cognitive impairment. He also had mild adult-onset diabetes and wasn't very good about monitoring his glucose. His wife, in contrast, was all over it and all over him—the matter was clearly a source of ongoing friction. He had also had a total hip replacement in New York, the result of an injury in his adolescence that finally caused enough arthritis to require a repair. He had been a smoker ("You would, too, Doctor, if you were worried about the nuclear weapons in Kazakhstan"). He wasn't much of a drinker. He loathed exercise; he said it was for "kids, and he hadn't been a kid for forty years."

He seemed to be compliant with the significant number of medications he took. He took three pills for hypertension, two for diabetes, aspirin, and a cholesterol-lowering agent. It is no mean feat to have a patient take this many medications, all in the hope of preventing further illness. Many patients, even if they can afford the cost, simply skimp on the amount of medication they comply with, some because they think physicians are worrywarts and like to overprescribe, and others because they can't be bothered and don't appreciate the studies that support the use of the medications. And, of course, we doctors rarely take the time to go over, pill by pill, the rationale for each and every prescription. This is further compounded by the reality that many of the pills the patient takes are prescribed by specialist physicians. It's often left to the primary care doctor, at each visit, to review the medication list and revise it. I was pretty good at doing that, but I almost never went over the rationale behind each medication, no matter who had prescribed it—there simply is not enough time in the day. Electronic medical records now enable the doctor to see when a medication has been refilled—an indicator of compliance—but that, too, is time-consuming, and I only did that when I started to get suspicious about a patient's compliance. There is silent but enormous pressure to see many patients in each half-day session, and doctors are usually silently triaging how much ground to cover in each patient's visit.

My first examination of W.S. revealed him to be quite bright (if he had lost some cognitive function during his second bypass, he would have been intellectually most formidable before it) and with a dry sense of humor. He had signs of disease in the blood vessels seen in the retina, which can be surrogates for the state of the coronary arteries. He had strong lungs and lung function. He had no liver abnormalities, but he had mild damage to his peripheral nerves, something often seen in people with diabetes. He had a dark lesion on his mid-back that had a surrounding region of depigmentation. This is sometimes a tip-off of a malignant melanoma. We chatted about how I ran my practice, about my partners and my coverage arrangements, and about navigating around the hospital, both literally and metaphorically. I told him I thought he needed a few cardiac studies and that it probably would be wise to see a cardiologist on-site, in case of problems down the road, and he definitely needed to see a dermatologist.

He did all I asked, and he really liked his dermatologist. The dermatologist is a brilliant guy, but definitely different. He loves hats, and I would often see him entering the hospital wearing a Sherlock Holmes–style deerstalker hat and cape. This same dermatologist confirmed that the back lesion was a malignant melanoma, and he did plastic surgery to excise it. Melanoma is a tricky cancer, but there are some facts that can give a patient guidance. First, if the melanoma is not very deep, that is a good thing. Second, the cell type can be on a spectrum of being more or less aggressive. And last, the surgeon looks at local lymph nodes when removing a tumor, and if there are no melanoma cells in the lymph nodes, that is also a good sign. But this tumor tends toward micrometastasis (spread that is visible only under the microscope), and that spread can light up, years later, often in the brain.

W.S. took the diagnosis well, and thought I was one clever fellow, picking up the melanoma on a first visit. If a patient thinks you are a good doctor, more than half of any further problem is likely to be solved because there is a foundation of credibility.

It turned out that the melanoma was not an issue. Over time, his prior hip replacement on one side became increasingly troublesome. He was a tough fellow and could live with the nocturnal pain it was causing, but the hip was also slipping out of its socket, and that was a real problem. So I sent him to one of our gifted orthopedists who specialized in redos; that is, he was an expert at fixing hip replacements that had either gone bad many years down the road or had never been a good result from the start.

W.S. was seen by appropriate consultants prior to the surgery, and on the day of the operation, he was optimistic and prepared mentally and medically. The surgery went well, and within three days he was at a rehab facility. Within ten days he was home, continuing his physical therapy.

Like many men his age, he had some problems with voiding. Usually this is from an enlarged prostate gland. The prostate encircles the urethra, the tube out of which men urinate, and with the passage of time, the prostate enlarges. When it enlarges, it can cause a blockage that needs to be surmounted to enable the passage of urine. This blockage can make urination difficult; it can make the urine stream weak or barely present, and it can lead to a lot of dribbling after the act of urination. It's a common problem in my practice population. Medications are used, often initially, to either enhance the power of the stream or to reduce the size of the prostate gland over time. W.S. refused the medications that were discussed, so I sent him to a urologist who had been a colleague of mine for years. They made arrangements for surgery on the prostate gland, a procedure in which the prostate is essentially cored out from the inside. This procedure is called a TURP (a transurethral resection of the prostate), and it is a very common procedure. W.S. did well and went home three days later.

Whenever surgery is done, the tissue is submitted to a pathologist. And, as is sometimes the case with prostate surgery, a bit of the tissue showed cancer. What to do about prostate cancer is an active quandary for both physicians and their informed patients. Pathologists, reading the tissue, grade the tissue according to a Gleason score—a two-part additive score that has weak correlation with the tumor's behavior, but at present it's the best we have. In the past, surgery was recommended for patients with high Gleason scores. Currently, if the cancer is spreading beyond the prostate, surgery is still recommended, but that, too, may change as long-term studies become available. Some recent data from Scandinavia (where registries follow citizens for years and where surgery for prostate cancer is rarely done) show that watchful waiting is often less morbid for the patient than radical surgery. Of course, absent hard and fast data, in the United States the patient often wants to attack the cancer in the belief that doing something is better than doing nothing. It is to be hoped that improvements in diagnosing which cancers are likely to be aggressive will help us decide who is best suited for surgery and who would not profit from surgical intervention.

But these data were not available when W.S. was told he has a Gleason score of 9 (out of a possible 10), and his urologist, who is very capable, did suggest surgery, a radical prostatectomy. W.S. and his wife readily signed on. And that's when the trouble really started.

A major problem in hospitals is the presence of certain bacteria that lurk in the corners of rooms, in the ventilation systems, and on the hands and clothing of hospital workers. These bugs have been around for a long time, and they are case-hardened and malevolent actors. If you get them into your bloodstream, hang on, because you are in for a hard ride.

W.S.'s radical prostate surgery went smoothly. I saw him in the post-op unit when he was waking up and looked none the worse for wear. A few hours later, he was routinely transferred to the surgical floor. When I saw him the next morning, it was clear he was in trouble. The night of the surgery, he spiked a fever at 104°F, and his blood pressure dropped. The surgical resident covering started broad-spectrum antibiotics and fluids, but it wasn't enough. W.S. was growing a strong strain of an antibiotic-resistant staphylococcus, and it was seeding his body. Staph has a predilection to stick to tissues, and particularly to stick to any foreign matter in the body. W.S. had a metallic hip implant and an artificial valve, and his staph stuck on both.

Having presided over far too many patients with staph sepsis (sepsis happens when bacteria invade the bloodstream and cause multiple derangements, such as circulatory collapse), I arranged to have the patient urgently transferred to our medical intensive care unit (MICU). Our surgical colleagues are happy to have us manage infectious complications, so there were no political issues regarding his transfer.

The issue was threefold. First, would W.S. survive this bout without succumbing to the septicemia (bacteria in the bloodstream) caused by the staph infection? Second, if he did survive, what complications would ensue? And third, what was the cause of the staph sepsis?

He did survive, but just barely. For two more days, the staph seeded his bloodstream. His body's response, a high fever and high pulse rate, caused a large heart attack; this caused further shock and required not just fluids but artificial support of his blood pressure with medications (known as pressors). During this chain of events (heart attack, shock, use of pressor medications), his kidneys shut down. All of these are common scenarios with severe sepsis.

With support and superb nursing care, he started to rally. All of his team members (residents, MICU attendings, myself) were examining him twice daily, and on day 5 it was clear he had a new cardiac murmur. An echocardiogram confirmed the presence of an excrescence on his aortic valve—a vegetation, in medical parlance. It was clear that the vegetation was causing the valve to leak and would likely require replacement. But he was too unwell to go forward with risky cardiac valve surgery now.

His kidneys started to work again, though with some permanent damage, and medications were given to try to control the heart failure induced by the leaking aortic valve, when he started to spike daily fevers again. Blood cultures were again positive for staph, and the team turned their consideration to his artificial hip prosthesis. We know that staph often seeds, and within a week or so, evidence of abscess formation becomes apparent. An ultrasound of his hip demonstrated a large fluid collection in the hip joint. An orthopedic colleague tapped the fluid and, sure enough, it grew staph.

So, to summarize, poor W.S. went in for routine hip surgery, had urinary difficulties afterward, and ended up with his prostate removed. But much more seriously, he endured hospital-acquired staph sepsis with compromise of his aortic valve and his artificial hip. This is an all-too-common scenario. A minor, routine matter escalates after a hospitalization, leading to more and more serious consequences and procedures.

After a number of conversations with our team and excellent infectious disease consultative help, it was thought that the initial TURP procedure somehow instilled resistant hospital staph into the prostatic bed (the area beneath the bladder where the prostate is located), and when the surgeons later went in to do the radical prostatectomy, they unknowingly seeded his bloodstream with antibiotic-resistant staph.

As the primary care doctor, my job was to explain the necessary next major steps. I needed to make certain the patient and his wife understood the gravity of the situation, the long time frame for potential recovery, and the risks of all further surgical interventions. Once recovered from his severe sepsis, W.S. would need cardiac valve surgery, and then, once recovered from that procedure, he would need a new hip. I explained how a quick procedure would be done in short order to remove his infected hip and to put in a sterile block where the hip ball joint was. He wouldn't be able to walk on the hip until he had a new one, but all of the problems, though significant and surgical, could be fixed. It would be a long haul, with ups and downs, and would require intensive rehabilitation while he continued on antibiotics and received nutritional replacement.

I coordinated a team meeting with his wife, the nursing staff, a social worker, the orthopedist, the urologist, the cardiac surgeon, and the cardiologist called in on the case. All of the family's questions were addressed openly and without gilding the lily.

Not surprisingly, W.S. and his wife needed time to process all of this. The ability to give patients and their families time has been compromised over the last few years, as length of stay in a hospital is an important statistic that has become ever shortened, giving patients and their families less time for recovery and decision making. W.S. had to be transferred to a rehab facility and needed two visits by me (at the end of some full days) to fully discuss all his options. Because this messy situation was a complication of his surgery, the family considered having W.S. receive his care elsewhere, and I spent some time introducing them to colleagues at other medical centers in Boston.

Ultimately, all went well. W.S.'s aortic valve replacement occurred six weeks after his sepsis, and he had no further serious issues prior to the valve replacement. Then, two months later, he had a new prosthetic hip placed. It turned out that both procedures were done in our institution. And it was truly one of the happiest days in my practice when I saw him walk into my office, unaided, and say, "Well, doctor, I made it!"

FRIDAY NIGHT AT FIVE

LIKE MANY OF MY PATIENTS, E.M. WORKED IN the hospital. She was a psychiatric social worker who worked specifically with childhood cancer survivors now being seen as adult patients.

Large hospitals have employees who often have very specific preferences in physicians, and word spreads quickly as to which physician is good at what. This patient had first been seen by a partner of mine, with whom she didn't get along. Physicians are often wary of sophisticated medical consumers, as they can be demanding and often know how to work the system. Furthermore, in a group practice, covering partners get to know each other's patients well. This patient had left one of my favorite partners, one with whom I had many discussions about movies, books, and life in general. I was concerned that a patient who felt incompatible with my partner, someone I felt I had much in common with, would feel incompatible with me as well.

So it was with more than a bit of trepidation that I saw her in my office. E.M.'s problems were not major; she was a woman in her late forties, menopausal, and burned out at work. She thought menopause might be causing some memory problems and wanted to pursue that. She was going to discuss protecting her bone health going forward with her gynecologist, and she was not having any sexual difficulties from her menopause. She had a happy and solid marriage and two children, both nearly out of college. A sister had breast cancer, but she was the only family member with any cancer history. Her brother-in-law, a famous orthopedic surgeon in town, had a bombastic and ego-syntonic personality, a point worth noting. Physicians who are part of a patient's family can make the patient's situation more complex since they are readily available to critique a primary care physician's judgment.

E.M. had a bohemian look that had left Boston around 1978, but she affected it: tie-dyed Himalayan hair bands, colorful Mexican shirts, and bright-colored puffy cotton pants. She took no medications except for some herbal supplements. She exercised regularly and took no illicit drugs. She used to drink in moderation years earlier, but had stopped using alcohol. She did regular yoga and Pilates. She did complain of some heartburn symptoms, and she had tried some over-the-counter preparations with good relief. The heartburn was intermittent and recently had become more severe.

Her examination was entirely benign, and she looked younger than her forty-nine years. I was pleased at the end of the encounter to not have any misgivings about her transfer to me, as she seemed genuinely just not to have hit it off with my partner, and trust and confidence in the doctor-patient relationship are very important.

We decided to see if we could eliminate her heartburn symptoms completely with a two-week course of omeprazole (an over-the-counter acid suppressant). Should that approach not work, I proposed that a gastroenterologist look at her upper gastrointestinal tract with an endoscope to ensure that all was well there. I ordered a test for *Helicobacter pylori* bacteria and, though the test is not a perfect one (it has a high rate of false negatives), she did not test positive for these bacteria. If infested with these bacteria, patients may have refractory heartburn, unless the bacteria are treated with a combination of antibiotics plus omeprazole.

E.M. took the omeprazole and called me in two weeks to tell me that it seemed to have eliminated her symptoms. I told her to call me in a month to report in and to see if she was still symptom free. When she called approximately six weeks later, she was feeling fine. We reviewed her lab tests from her examinations, and she told me about the plans she made with her gynecologist for checking her bone health. I didn't hear from her again for seven months.

The worst time to call a doctor's office is at 4:50 P.M. Assuming that office hours end at 5 P.M., most physicians are trying to triage the end of the day. They are making all the required return phone calls, checking in with their spouses to see if they will make it to dinner on time, considering whether to call the house staff caring for their hospitalized patients, and thinking about maybe even getting up and stretching during that focused hour or so before they can shut off the lights and turn off the computer.

So it is that almost all doctors' hearts sink when they get that call at 4:50 P.M. Even worse, of course, is 4:50 P.M. on a Friday afternoon, when the desire to get out is sharpest. Often these calls are prefaced by "I know it's nearly the weekend," and then patients relate that they have had back pain for six weeks, or they have been coughing up blood for two weeks, and that they thought they'd call in case it was important and the doctor would not be available for the next two days. I think few physicians ever get over the frustration of those late Friday moments.

E.M. called late on a Friday afternoon and said, "I think it's nothing, but my heartburn is back." I, too, thought it was nothing but did the usual: I asked when the symptoms started, if they were exactly similar to previous symptoms, what provoked the heartburn, and if she had tried to relieve them.

It became clear that this heartburn was a little different. E.M. could not articulate exactly how it was different, but it was. She could state with certainty that the discomfort was more intense and that there was more of a burning feeling than with her usual heartburn. It seemed to happen mostly when she was exercising, but over the last three days, it had been happening with very little activity. There was something going on here that was hard to quantify. This woman would not have called (especially at this hour) if she were not truly concerned, and her original heartburn had not concerned her very much at all. Also, this sensation was more intense and newly associated with her regular exercise.

I told E.M. that her symptoms concerned me. I recommended that she go to the emergency room and that I would meet her there. Like most patients do when the two magic words "emergency room" are introduced into the conversation, she gave many reasons why this was simply impossible: She had private patients scheduled for this evening, the wait in the ER was always long, she was concerned people would know her and think her foolish to be there, and so on.

I have heard all the excuses, and I am not one to pull the trigger lightly on an ER visit. Those visits are expensive, often the patient has a large co-pay, and the scene at the ER is often one of pandemonium. Furthermore, I seriously embraced the role of a primary care physician in heading off needless ER visits as a proper cost-saving measure for all concerned and to help unclog ERs.

But E.M.'s symptoms niggled at me. I thought it unlikely she had anything beyond a recurrence of heartburn, but the intensity of the pain, her call that disclosed her concern, and the relationship of the pain to exertion were all red flags. These were balanced by her relative youth, lack of any family history of heart or vascular problems, and her good health habits. As with so many efforts to do the right thing for patients, we negotiated. She wanted to be sure that I would be there in the ER to move things along and that she could have my opinion, along with that of the ER attending. If all was well, she wanted to be able to leave promptly to keep her evening client appointments.

I was blessed with having an office at the hospital, so to go to our ER only required walking a few steps. Because of this, our ER attendings knew that we saw our own patients, and they appreciated it because it expedited things for them. Of course, our patients appreciated seeing a familiar face in that stressful environment—as other surrounding patients could be seizing, screaming, or having cardiac arrests. And we could often expedite patient care by knowing what to pursue and what to ignore in the evaluation of the patient.

E.M. looked anxious but comfortable in the emergency room. Her blood pressure was a little elevated, but that is very common with the first set of vital signs in the ER; the vast majority of people are anxious. Her lungs were clear, and her pulse was regular. Her cardiac exam was unchanged from prior examinations in the office and showed no signs of a problem. However, her electrocardiogram (EKG) was subtly abnormal and showed some damage in the back of the heart (we call this the posterior region). The findings of a true posterior coronary are often subtle, and I had the advantage of having her EKG from a few years earlier.

She had had a heart attack. The question was, When? She was given potent blood thinners and taken urgently to the catheterization lab—something we can do in our hospital even in the middle of the night. Her coronary arteries were quite clean and showed no significant atherosclerosis. So what happened?

The honest answer is that none of us truly knows. We do know she has no significant atherosclerosis. But did she have the heart attack from stress? Was it from spasm of the artery that supplies the area? Was it from a mismatch between the demands of her heart muscle and the small blood vessels that supply those muscle fibers? Her case highlights much of what we do not know about cardiac function and damage. It further highlights that we have done far fewer studies of women than of men in medical research. Consequently, the database available for women with cardiac issues is much smaller. This national problem is being addressed, but studies on the natural history of cardiac disease take a long time to complete, so the database will remain meager for some time yet.

But here, the doctor's dilemma is how to counsel the patient. I had to tell her she had a heart attack, though it had been small. And I had to tell her that I didn't know why she'd had the heart attack. I knew she would seek answers, so I came prepared with the name of a female colleague who was a preventive cardiologist with whom she could consult. Two days after admission she was discharged with medications we know can prevent a second heart attack and can prevent dangerous remodeling of the heart after it is scarred.

E.M. likes her cardiologist, who has reinforced all the good things the she had been doing prior to her myocardial infarction and has encouraged E.M. to leave her job to find something less stressful. The cardiologist has shared her considerable knowledge of what is known about women in midlife who experience cardiac problems, and this depth of knowledge, I suspect, has been more therapeutic than any of the other interventions.

LEARNING FROM THE PATIENT

FIRST YEAR OF MEDICAL SCHOOL WAS TORTURE, but Second Year was fun. Things have changed a lot in the more than thirty-five years since I graduated, and if they continue to change at such a rapid pace, medical students might actually be doing complex neurosurgery during their first two weeks of school, in the name of keeping their education pragmatic and patently useful.

But back when I went to medical school, our education was not so sped up. First Year was devoted to normal systems, and Second Year focused mostly on the abnormal. Who, in their right mind, doesn't prefer abnormal, really?

Studying normal systems involved anatomy, and my classmates and I spent hours and hours in a smelly room with a bizarre chap who supervised the handling of cadavers. He was the deinar—the keeper of the cadaver room—and he always looked a bit unwell or worn out. He was around five feet tall, wizened, was always sneaking outside for a smoke, and had "Lone Hawk" tattooed on eight of his fingers just below the knuckles. He was supposed to keep us from throwing cadavers' ears around the room or tossing livers in the air like pizza dough. As distasteful as anatomy lab experience was, we then had hours upon hours of biochemistry. Memorize this pathway, then the next, then the next—all really without any context to put them in. Then we hit the microscopes to look at normal cells. I could tell you the difference between a thyroid cell and a testes cell in no time. Next, we would trot off to the microbiology lab to learn about all the bugs that can do you in. I actually liked this course, as it was full of colorful bacteria (both their reputation as killer bugs and their color under the microscope when stained)—but it was ruined by a lab instructor who was so depressed that no amount of enthusiasm expressed by anyone could bring him out of his deep depressive state. We also learned physiology—the normal functioning of all the organ systems—and at least in this course we could see the light; we realized that, at some point, we were going to apply this knowledge. The one downside to this course was that we had to experiment and prove concepts on animals. Anyhow, you get a picture of First Year.

Second Year was entirely fun. Building on physiology, we learned what could go wrong to screw the body up. We learned pharmacology—how drugs are constructed, absorbed, and eliminated and where in the body they acted. We also learned about the nervous system—how amazing it is when things work properly, and how utterly awful it is when the wiring messes up. We also had a course in tropical medicine taught by a popular teacher. The topic itself is fun—how those little parasites find the human body a delectable host—but it was made even more delightful by the raconteur who taught it. When talking about an amoebic infestation, he embellished a tale of being in Manila with a very pretty woman whom he was entertaining when he was stricken with "loose movements" from amoebiasis that would not quit. And when talking about a tapeworm, he described a patient, a very famous and buxom movie star, who had passed this worm and brought it, in hysterics, to his office. His reputation was cemented further by the fact that he smoked a cigar and drank Pimm's Cup cocktails while teaching the course. Many of us brought our significant others to his classes, as he was one of those people who seemed unbelievable until he was seen in action. It wasn't often that you had a lecturer get half-loaded while ostensibly teaching a class about tapeworms and other wriggling critters. And if that wasn't enough fun in Second Year, we finally got a chance to examine patients.

The introduction to patients was truly exciting. Now, those of us who were interested in this field of medicine—namely, practicing and being engaged with patients, as opposed to being on the lab bench for a lifetime of research—could see why we had endured the torture of First Year. We were tutored in ethical behavior toward patients, taught how to interview patients in a sensitive but probing manner, and taught how to use the tools of the trade. We were instructed in the use of the stethoscope, ophthalmoscope, tuning fork, and reflex hammer—all things I use, with both pleasure and pride, to this day.

The purpose of those first two years of medical school is to teach us how to integrate the normal with the abnormal and, thereby, how to understand the underlying problems of our patients. With that knowledge, we can consider what might be a rational approach to manipulating a patient's underlying problem so we can achieve an improved outcome for the patient. With heart failure, this approach has been moderately successful. Drugs have been developed that capitalize on the known dysfunctional aspects of the heart, and, by manipulating their application, we have improved patients' symptoms (though not yet outcomes) of this common ailment.

One of my patients, Mrs. T.C., had a condition that derived from something I had studied during my first year of medical school—but at first I didn't recognize it. When I did finally understand that her symptoms were caused by a change in the physiology I had learned about so long ago, I went back to the books to refresh my memory. One of the reasons I didn't recognize the situation was that many of the studies we did then were performed in a dog lab, where our experiments were done to teach us basic physiologic processes. However, the human body is not the same as a dog's: Humans are much more complex and, truth be told, uniquely human. When I went back to basics to understand Mrs. T.C.'s condition, I realized how incomplete my studies then were and even how incomplete they remain today.

Mrs. T.C. was referred to me by a friend and patient. She had been seen by many doctors in the last few years, and she felt that no one had given her a unifying diagnosis. More crushingly, she felt she had been dismissed as a whiner and complainer, and she believed that no doctor could understand that her condition was causing her increasing distress. Her symptoms had also been increasing: weight loss, depression, anhedonia (lack of enthusiasm for life, sex, fun, etc.), palpitations, headaches, light sensitivity, cold extremities, confusion, and multiple food sensitivities.

When I first met her, she was a very slender, clearly capable woman who had had a career as a social worker. However, about a year earlier, she was forced to give up working because she felt she could no longer concentrate on her job. Her inability to function at work seemed to be twofold. First, she was not thinking well. And second, she would have overwhelming symptoms, such as palpitations or headaches, that would be completely distracting. Fortunately, her husband was a physician, and financially she could afford to take an early retirement. But she truly enjoyed her work, which gave her both structure and meaning that she did not want to relinquish. For her, the salary was secondary. She related this to me with considerable crying and with the stress and frustration of meeting yet another of many physicians. As she related all this to me in our initial encounter, she started to experience palpitations and a severe headache. I didn't have to be told to immediately interrupt our interview and take her blood pressure, which was very elevated (210/115). Her cardiac impulse was also strong and pounding.

I asked her if she felt we could continue the interview, and she asked if I would leave her alone, with the lights on "as brightly" as I could make them, and come back in thirty minutes. This was an unusual request, and it is one of the sad aspects of medical practice today that doctors are routinely booked, usually by a computer program, for a patient visit every ten to twenty minutes. Just to meet Mrs. T.C.'s request, I was going to fall far behind in my morning schedule. But my sense was that there was something genuinely organic going on, and, even if turned out that the cause of Mrs. T.C.'s condition were psychiatric, she would have then been the most fascinating patient with a pure psychiatric disease that I would have encountered in years. I agreed to her request, left her alone in an exam room, and found and apologized to the two waiting patients who would have to be delayed or rescheduled.

When I returned to the exam room, Mrs. T.C. was calm, although frustrated and tearful, and her blood pressure had become low normal, something that I had originally expected with her slender frame. She told me that this situation—sudden headache and elevated blood pressure—was an everyday occurrence, sometimes happening two or three times a day, and that she had been dismissed by many physicians as crazy or melodramatic, and that even her husband was becoming quite frustrated with her. It was clear, even without her elaborating, that she could not do much outside of her home, as these attacks were disabling.

I asked her how much more of an intake interview she thought she could tolerate, and she told me to go ahead and try to glean more information from her because, as she explained, the very acts of getting dressed, putting on makeup, and negotiating Boston traffic were efforts she did not want to go to waste.

I was then able to ascertain that she had had a benign medical history until ten years ago. She had felt a firm bump on the right side of her neck, which was investigated and found to be an unusual tumor in the nested nerve tissue in the carotid body. Its formal name was a paraganglioma, and it was treated by excision of the junction of the internal and external carotid artery. This region of the neck contains the carotid sinus and carotid body, which have a number of functions, but the one I was concerned with is the input they provide to the very basic, old part of the brain, the medulla, for blood pressure regulation. The medulla controls activities like keeping your blood pressure intact, and it, and a region slightly north of it, regulates your breathing. Both blood pressure and breathing have inputs to the medulla. The carotid body region sends a message to the medulla tonically—that is, all the time. When blood pressure goes up, more nerve impulses are sent to the medulla, which, by the nerves exiting it, slows the heart rate and reduces the strength of the heart's contractions. In this manner, blood pressure elevations are mitigated and sent more toward normal pressure. I'm not talking about the phenomenon of essential hypertension here but, rather, acute changes in blood pressure. So, if this region is damaged or, as in Mrs. T.C.'s case, excised, that tonic input to the medulla is gone. Brains don't like being disinhibited like this, and they become volatile. Blood pressure swings tend to be wide and not autoregulated promptly. A light bulb in my brain went off: Maybe some of what Mrs. T.C. was experiencing was related to the physiology I had learned in my first year of medical school so many years ago.

She told me that since her surgery she had had some blood pressure lability, but that over the last eighteen months it had gotten much worse. We didn't have time for too much more that day. I examined her, did an EKG, and asked her to send me her records. I suspected that some of her prior physicians had diligently done some thorough and thoughtful investigations. Mrs. T.C. revealed that she really was interviewing me to see if I'd be a suitable physician for her to work with, and she had to think about it. I told her that was fine with me, that she should make another appointment, and if she was okay with us going forward, to get me the records so I could read them prior to our next visit.

She did come back, and I read her records. She had had a number of capable and sympathetic physicians who had evaluated her for an unusual tumor called pheochromocytoma—something I myself would also have done. As pheochromocytomas can have intermittent symptoms, it is worth evaluating a patient for this a few times to catch the tumor when it is active. Furthermore, when a patient feels stress, the fight-or-flight hormones are released. These hormones are the same as those released by the pheochromocytoma, so the results can be confusing. But her pheochromocytoma evaluations, on more than two occasions, were normal. Her records for these investigations were complete and thorough. That hypothesis had to be discarded.

On her second visit, I took a more detailed history. Nothing was striking. Her physical examination, though, revealed a lump in her left carotid body, which was beyond interesting. Could a second tumor be the cause of her increasing symptoms? A brief work-up ensued, which confirmed the presence of a second tumor. This time, the new tumor was treated with radiation. As expected, the now total obliteration of both of her carotid bodies did not help her symptoms; rather, it exacerbated them. But this had all been discussed with both her and her husband when the second tumor was found.

Here we had a woman, trapped by the disinhibition of one of her blood pressure regulators but who had many other symptoms besides the expected wide blood pressure swings. Many of these symptoms were unexpected; why was she sensitive to darkness, experiencing mood swings, and suffering from food sensitivities?

The next step was the most crucial. I told her that I was puzzled by many of her symptoms but that I was going to do some research. Over a few hours during my evenings at home, I searched through the medical literature widely available on the Internet. I found there was one physician in the United States who had published on patients with carotid body tumors similar to Mrs. T.C.'s. I called him one afternoon and described her symptoms. He told me he had by then seen a handful more patients than those he had originally published about, and a few of them had symptoms that overlapped with my patient's. And, he said, he would be happy to see her. I was delighted that he had made the time to take my call and enthusiastic that he had seen other patients with symptoms very similar to Mrs. T.C.'s. I called her at home and shared the result of my investigation. She was more than eager to see this expert and his team, with the understanding that she might become a subject for further investigation because of her arcane condition.

Mrs. T.C. still has many of her symptoms. Her visit to the expert did not yield a cure. But, most important, it validated her symptoms, gave her the perspective that others had suffered in a like manner, provided her with the experienced physician's ongoing counsel and support, and gave her hope, as the expert promised to share any knowledge he learned with her.

Today Mrs. T.C. does a lot of transcendental meditation, takes some chronic mild blood pressure medications, and supplements them with both anxiety-inhibiting and rapid-acting blood pressure medicine as needed. She winters in the sunny

Southwest and returns to Boston late in the spring when the sun is strong. She is not cured, but she is empowered.

For me, the lessons were considerable. I had not thought about the carotid body for some time, and what I had learned so many years earlier was only a small part of what is known now, as the experiments supporting knowledge of the carotid body back then were done on dogs, not humans. I was reminded that searching for answers when I am puzzled is the right thing to do—and it can really help a patient. I was reminded that a patient could become her own best advocate, in this case one who tells me new things she has learned over time from her expert physician counselor, from her own blogging, and her own self-experimentation. One fact she discovered is that her rare bilateral tumors have now been associated with an aberrant gene, and she has encouraged some of her relatives to be tested for it. And of course, the final lesson, for both doctor and patient, is to never give up trying to find the answer.

EXPLOSIVE ILLNESSES DO NOT
RESPOND TO HOMEOPATHY

SHE WAS A SLIGHT, RESERVED AFRICAN American woman whose kidneys had stopped working. She had been urgently admitted from the ER to our inpatient service and had been feeling poorly at home for about a month. She had decided to cleanse herself with various homeopathic agents, but she noticed that she wasn't feeling any better and had developed a cough that would not stop. When she started coughing up blood, she came to the ER.

She lived with her two granddaughters in a nice middle-class neighborhood. Her granddaughters were identical twins and college track stars. She was very proud of them but was quick to articulate that she didn't like the music they listened to or their manner of dressing. She had entrenched opinions about how much care she wanted and how we doctors sometimes interfered with "the good Lord's plans."

She was retired. She had previously worked as an administrator to physicians in two other teaching hospitals in town. She was independent in her living and had devoted children who saw her most weekends. She wasn't well-to-do; she was paying her living expenses and still had a little left over for emergencies, but there was no disposable income for fun.

On examination, she was small and frail. She looked chronically unwell. She had a low-grade fever (101.3°F), a rapid pulse at 116, and a blood pressure of 180/110 (quite elevated). She showed signs of pneumonia in both lungs, and her cardiac examination showed signs of long-standing hypertension and some congestion. Her abdominal exam was benign. Her neurologic examination showed what appeared to be a small stroke on her left side—she had slight arm and leg weakness and slightly increased muscle tone on that side. She was too weak to get out of the bed and walk for us, but she had managed to take the streetcar to the hospital (which required a bus-to-trolley transfer) some six hours earlier.

Her laboratory values were markedly abnormal. She was very anemic, had an elevated white blood cell count, and had little kidney function left. She had a lot of abnormalities in her blood as a result of her kidneys not doing their job. Her chest X-ray showed infiltrates in both lungs, and her blood oxygenation on room air was moderately diminished.

Our team looked at her urine for clues as to what might have damaged her kidneys. Her urine showed signs of inflammation in the kidney filters (called the glomeruli), and we saw signs of more generalized inflammation in her kidneys.

She was urgently ill, and, truth be told, in many other hospitals she would have been managed in the medical intensive care unit. But in our hospital, you need to be desperately ill to be assigned to one of those limited number of beds, and she was only seriously ill, not desperately so.

She needed dialysis, and promptly. That would help her general feeling of malaise and her sense of being globally unwell. Dialysis would remove some of the toxins in her blood and would remove some of the extra fluid that the kidneys normally would not have permitted to build up. We also needed to gently lower her blood pressure. And, of course, we needed to figure out the cause of her underlying illness.

Medical doctors are used to pattern recognition. One of my colleagues calls it "airplane spotting"—that is, if you see an Airbus 320 it looks only like an Airbus 320, and it looks different from a Boeing 787. Rapid pattern recognition is a plus for an internist. It favors people whose minds make cross-references readily. It is a skill that can be taught, but some people are naturals at it.

My resident was a natural. She had demonstrated excellent command of the management of the diverse patients on our floor. She was also a wonderful team leader and naturally and spontaneously reached out to all members of the care team for their input to our patients' problems. But where she most excelled was in seeing what the diagnosis might be given disparate pieces of information.

She told me, "I believe this is a pulmonary-renal syndrome and that the infiltrates in the lungs and the kidney damage are related." She wanted to explore this leading diagnostic possibility with appropriate blood testing. I thought she was correct. It is my job, with a young doctor-in-training, to support her diagnosis when it is on the mark and to guide her to a better route when it is off the mark. Furthermore, with gifted residents, ones who usually jump to the correct diagnosis readily, one of my jobs is to rein in their tendency to cast their net too widely. I asked her, "If this is what you think is going on, what three tests, and only three tests, should we order?" We conferred for a while and decided on an anti-glomerular basement membrane antibody, a p-ANCA, and a c-ANCA. These three tests all look for unifying diagnoses that involve simultaneous organ damage to both the kidneys and the lungs.

The lab tests were sent off. The kidney (renal) service was called to consult and, because they control use of the dialysis machines, to get the patient dialyzed. The rheumatology service was also called, as the diseases we were considering are seen by both renal and rheumatology. And, because of our patient's hypertension and heart failure, we ordered various cardiac tests.

The dialysis occurred without incident, and the patient felt much stronger and mentally improved. She told our team that she had had kidney failure once before at a different teaching hospital in the city. She had been treated with high-dose steroids, which she had despised because they gave her insomnia and made her legs and arms weaker. She had made a complete recovery, but, because she felt the doctors there were disrespectful, she stopped going there for follow-up around two years ago. She also told us that twenty years before she had had a seizure and an extensive neurologic workup at that time, which included a brain biopsy (not something that is done lightly) that showed inflammation of her brain blood vessels. She was also treated with high-dose steroids at that time and never had another seizure. She stopped having follow-ups at that hospital because it was where she worked as an administrator, and she felt that too many nosy people had heard about her illness.

Her illness is called vasculitis—inflammation of the blood vessels. But the term does not do justice to the various conditions lumped together in this diagnostic bucket. Some types of vasculitis just involve tiny veins in the skin and can cause little raised skin bumps. Other vasculidites (these kinds of blood vessel inflammations) are like the ones our patient had—if not diagnosed and properly managed, they will surely kill. Some are indolent (slow to develop) illnesses, others explosive (they develop quickly)—as hers had. Some involve tiny veins, and others involve the largest arteries in the body.

These conditions, the body of illnesses that constitute the vasculidites, have fascinated me since medical school. I saw my first case as a third-year student on my medicine rotation. The patient was a man with inflammation of the arteries that nourish the intestines. The consultant rheumatologist who was called was a handsome, thoughtful professor of medicine and basic researcher. I told him I was the student assigned to the patient, and he was extremely kind to me, trying to help me gain an understanding of the condition and its ramifications for the patient. He was so gracious to me—a very green student—and so clear in his reasoning that I decided then and there that I would try to emulate him. Emulation is never 100 percent, and so I never managed to be as handsome, or as intelligent, but I did end up working in his laboratory as a fourth-year student and going into the same field, rheumatology.

This calm, gifted, erudite man, named Charles Christian, taught me many things. He role-modeled how to best interact with a frightened and very ill patient. I watched him lead his colleagues in his division (he was the division chief) and saw how he was keenly interested in their research, grant applications, upcoming publications, and varied personalities. His managerial style taught me a tremendous amount, a skill set I didn't even realize I was incorporating into my own thinking.

His manner with trainees, both fellows in his division and students, showed me how to be respectful to those who know little but who are genuinely striving to master a curriculum.

That patient I encountered as a third-year student had vasculitis of the medium-sized arteries. In his case, it was confined to the medium-sized arteries that nourished his intestines and kidneys. His kidneys, like those of my current patient, were inflamed. But, more urgently, a significant piece of his small intestine was compromised by an inflamed blood vessel and had to be removed. This is a serious surgical procedure and one that is high risk when the kidneys are not functioning. The drug armamentarium available back in those days was limited. The patient required and received high doses of steroids, which are not helpful for either wound healing or blood pressure. And in around ten days, in spite of the urgent and successful bowel surgery and ongoing dialysis, he died.

That patient left a strong impression on me—a curious, industrious, and very impressionable medical student. I had so many questions: How did they even make the diagnosis? Why did the treatment fail? What could his care team have done better? Did his wife really understand what his illness was and why he succumbed?

I determined immediately that I was going to try to master these questions. And of course, that is the arrogance of youth. Because, after more than thirty-five years in the business, vasculitis still fools and inspires me and holds out the promise of better diagnosis and therapy.

The reason vasculitis can be so occult is that it can affect blood vessels of all sizes. Remember, all living tissue has blood vessels near each individual cell. The blood vessels transport the oxygen that all living tissue requires and remove waste products. The oxygenated blood travels in arteries. The deoxygenated blood, which transports waste, is carried in the veins. Arteries and veins both are blood vessels, and vasculitis can affect both.

In the vast majority of cases, we do not know the inciting cause of the inflammation to a blood vessel. Rather, we see the consequence of the inflammation and try to put out the inflammation as quickly as we can because, if left unchecked, the inflamed blood vessels damage the tissue downstream of the inflammation, which can compromise the function of the kidneys, brain, heart, lungs, and skin.

Because blood vessels reach every part of the body, there can be multiple clinical manifestations of vasculitis. The patient might have a stroke, or cough up blood, or have sudden kidney failure, or experience abdominal pain and pass blood in the stool. The ability to suspect vasculitis is important; many clinicians do not see it often in their training, so they make the diagnosis late, or only when a consultant suggests it.

Often the vasculitis syndromes are classified by the size of the blood vessels involved. As a generality, very small end capillary venules present a vasculitic syndrome that is a palpable, purplish, disseminated skin rash. Medium arteries can affect the kidney or coronary arteries. Large arteries can affect or impede flow in the aorta or large aortic branches. Although there are certain blood markers that can be helpful in categorizing the type of vasculitis, there is considerable overlap in the utility of these markers, and, at the end of the day, extensive clinical experience is often the best diagnostic tool.

Now back to our current patient, who had had a seizure and had undergone a brain biopsy (a rather drastic intervention to make a diagnosis) some twenty-five years earlier, and then she got better. She didn't continue medical follow-ups after that episode. Then she had a bout of kidney failure around ten years ago and, once again, had no regular medical follow-ups. Now she was admitted to our hospital with kidney failure and coughing up blood. She also had very severe anemia, which we determined was caused by both the inflammation suppressing the function of her bone marrow and the amount of blood she was coughing from her lungs. She had an overlap type of vasculitis affecting mainly the medium-sized vessels, and we gave her high doses of steroids immediately. We also instituted an antibody to B lymphocytes called rituximab (it destroys B cells). Fortunately, both of the therapies started to work. Within two days, she stopped hemorrhaging into her lungs. Her kidneys took longer to recover—about ten days—and she had to undergo dialysis until they started to function properly on their own.

We consulted often with our colleagues in rheumatology and in renal. We all thought her prior hospitalizations were related to medium-vessel vasculitis. The likelihood of her illness recurring was high, though she tended to go for many years without any events. What allowed the illness to come to the fore remains a mystery to us all.

What about the patient and her understanding of her condition? She was quite intelligent. After getting to know her over her two-week hospitalization, I found out that, at the time of her first hospitalization, she was the main administrative assistant to the chairman of medicine at a prestigious Boston hospital. She knew a lot about medicine and academic medical centers. But she also was highly skeptical of the efficacy of medicine and felt strongly that prayer and homeopathic therapies were both better than traditional medical care. It became clear to me that she would probably follow up with her consultants in the outpatient setting for a while but then would stop all of her visits. I asked her if she would mind if I could check in with her by phone from time to time, and I was gratified that she said, "Yes, you may." More gratifying than her excellent response to her treatment when she was deathly ill was her declaration that she thought I was a doctor who cared. And because she felt that way, she said she would be honored if we could continue to speak to one another after her discharge. I still speak with her briefly about every two months, and she updates me on her grandchildren, her church, and, yes, that she is still doing well off all her medication.

SOMETIMES, ALL WE GET IS CLOSE

MRS. H. FELT FINE UNTIL HER FIRST PREGNANCY. Her early childhood was completely normal. In adolescence, though she felt she was overly concerned about her weight and appearance, she never indulged in any dietary excesses or severe strictures. She was never the "triple cheeseburger, fries, and chocolate shake" kind of person, but she ate her fair share of unhealthy foods.

She was an outstanding student and won an academic full scholarship to her prestigious state university. She excelled there and was a senior editor of the large and influential student newspaper. She achieved Phi Beta Kappa academic status while also holding down a work-study job and shouldering ever-increasing responsibility on the campus paper. And she met the man she would go on to marry and live with happily for the next thirty years.

Both she and her future husband stayed on at the university for law school. Soon after they graduated, they married, and Mrs. H. started a career in government service. After three years, she became happily pregnant with her first child. This was a planned pregnancy, and though conception was easy, the pregnancy was not. She often felt nauseous and excessively fatigued. She vomited a lot and lacked energy. But the worst symptom was a mental fogginess that prevented her from doing her job in an efficient way.

Her obstetrician thought this was just a first-time pregnancy in a high-achieving woman, and she kept reassuring Mrs. H. that all would be well. Indeed, she predicted that by the third trimester the nausea, vomiting, and mental fogginess would disappear. But she was wrong. One day, in the thirty-third week of her pregnancy, Mrs. H. was unresponsive, and her husband, alarmed, called 911.

Mrs. H. was admitted to the intensive care unit of a large university hospital. In the ICU, an extensive workup revealed that she had an elevated ammonia level. Ammonia, a compound with nitrogen in it, is toxic to the brain. It initially causes nausea, vomiting, decreased mentation, and, if it rises progressively higher, coma. Ammonia is generated in our bodies through normal biochemical reactions involving proteins. But as it is a toxic substance, it has to be removed from the body and converted to something else. In humans, this occurs in a biochemical cycle called the urea cycle. In this interconnected pathway, five enzymes need to be in good working order to remove ammonia from the body by converting it to urea. Urea is water-soluble and is readily excreted in the urine. If any one of the five enzymes is missing or deficient, the substrates for the enzymes are not normally processed and become backed up; eventually, they back up so much that ammonia levels in the body rise and become toxic.

As with many enzymatic pathways, full genetic absence of the enzyme is incompatible with life, and children die at a very young age if they lack it. However, particularly with partial deficiencies linked to the X chromosome—the sex chromosome that is usually doubled in women—much of the enzyme might still be present, and the urea cycle is then overwhelmed only under stressful circumstances.

Pregnancy is a just such a high-stress situation: The fetus develops its own circulation and growth, which places excessive demands on the mother. With normal enzyme levels, this is not a problem, but with partial enzyme deficiencies, a situation like this can tip the balance.

The doctors investigated the possible causes of Mrs. H.'s elevated ammonia. Liver failure, for example, is most commonly the cause in adults, as a lot of the urea-cycle conversion of ammonia to urea occurs in the liver. But that was not the case here. When all common adult causes of elevated ammonia were eliminated, the ICU doctors started to consider partial enzymatic deficiencies. All of the substrates of the five enzymes involved in the urea cycle process can be measured. In Mrs. H.'s case, it turned out that she was partially deficient in the enzyme ornithine transcarbamylase (OTC). Deficiency of that enzyme led to an increase in the two substrates that OTC normally converts and, thereby, caused a consequent increase in her ammonia level.

Because they are rather specialized, these tests took time. These kinds of tests require special handling and are done only in certain centers. Mrs. H.'s tests provided a disconcerting finding: When the actual enzyme was measured (and not the substrates that the enzyme acted upon), it was within normal limits. The doctors repeated the testing when Mrs. H. was clinically fine. And, again, the level of OTC was normal. Well after her delivery, her OTC was again measured, first when she was well, and again when she was deliberately sleep deprived for two days to stress her body. Again, on both of these occasions, it was normal.

How could we account for this puzzling finding? The explanation likely lies in the Lyon hypothesis, first articulated by the geneticist Mary Francis Lyon in 1961. She noticed that, even though females have two X chromosomes, one of them always seems to become inactive during the development of the embryo. This phenomenon was later modified in the 1990s by other investigators, who noticed that this inactivation was not 100 percent; the inactivated X chromosome became active 10–15 percent of the time. This could explain Mrs. H.'s situation. The scenario proposed would be that, most of the time, her healthy X chromosome with the normal genetic code for the OTC enzyme is working. But a small percentage of the time, her X chromosome with the deranged code for OTC is present. And in situations that cause a protein stress (because nitrogen, and hence ammonia, mostly comes from proteins), her small deficiency becomes evident.

Is this actually the case? Our current state of knowledge and instrumentation is inadequate to know for sure, but no one has a better proposal. As we do not know how to inactivate her low-level abnormal X chromosome, we have to tell her to avoid large protein loads—no huge steaks when having a celebratory dinner. Indeed, when she was questioned, she stated she had avoided excess protein since childhood—she just never seemed to like it.

Mrs. H. has generally done well. Since that first pregnancy, she has had only three other severe episodes. She is intelligent enough to know when she might be decompensating (when her body quits working as it should), and she has a medical alert bracelet that outlines what her treatment should be if she is incapable of telling the medical team herself.

It is conceivable that, in her lifetime, a way to inactivate that rogue X chromosome will be devised (and one hopes that, when it is deactivated, her problem will be solved without a new problem being created!). In the meantime, Mrs. H. continues to live a normal life. She is a walking example of medicine's ability to understand complex biochemical pathways but its inability to do much more that make a diagnosis—and not effect a permanent cure.

THINKING CAN SOMETIMES
MAKE A DIFFERENCE

D.J. HAD ALWAYS BEEN OBESE. AS A YOUNG girl, she hated gym period, as she was slow, heavy, and felt ashamed. She started and failed many dietary regimens into her forties. As she said to me, she knew more about calories and the values of certain foods than any doctor she had ever met—and certainly, when it came to me, she was spot on.

Physicians use a quick measure of obesity called the body mass index—the BMI. It has its problems as a metric, but it is widely used and provides a quick window into someone's weight.

The tables that life insurance companies utilize are very strict, so I always took their ideal BMI with a grain of salt. But, by any stretch, a BMI exceeding 30 is heavy, and one over 35 is quite heavy. When one's BMI is 40 or higher, it is obvious to loved ones, co-workers, and passersby that you are obese.

The other issue with a BMI over 40 is that it is often an indicator of other morbidities. Many people this large have adult-onset (type II) diabetes. Diabetes is bad news: It is a major cause of vascular disease (think heart attack and stroke), the chief cause of kidney failure in adults, and can cause debilitating neurologic symptoms, such as severe shooting pains in the extremities or lack of sensation that leads to amputations of digits or even legs. And if adult-onset diabetes isn't enough, people with BMIs over 40 often have sleep apnea (interruption of breathing, which can lead to hypertension and enlarged hearts), osteoarthritis of the knees, and liver disease (even progressing into cirrhosis). And there are a host of other illnesses, too, that are too numerous to mention. And, of course, in our youth-oriented society, it is the rare person who truly wishes to be obese—there is a stigma and shame associated with obesity that is likely as powerful a negative as the more easily defined medical issues.

Insurance plans recognize the hazards of this morbid obesity (which is what the condition of having a BMI over 40 is called) and will pay for bariatric surgery to address it. Although it can be done in a couple of different ways, bariatric surgery generally involves shrinking, tightening, or removing a good part of the stomach and attaching it lower down in the small bowel. This kind of surgery has become increasingly common in the United States over the last fifteen years, and now, over one hundred thousand such operations are done annually here. Most institutions that do the surgery have a multifaceted program in which patients first attempt dieting, receive psychological support and counseling, and are apprised fully of the risks and benefits of the surgery.

One of the benefits of bariatric surgery is a rapid reduction in abnormal glucose tolerance. It's a most interesting phenomenon—physicians know that significant weight loss reduces the resistance to insulin that obesity causes, but the improvement in glucose tolerance after bariatric surgery is rapid and occurs prior to any significant weight reduction. With weight loss, conditions such as infiltration of fat into the liver (which is what happens to geese being fattened for foie gras) improve, arthritis from excess load is ameliorated, and sleep apnea (due to obstruction of the airway from there being too much flesh around the upper airways) usually improves a great deal. But all these things take time and are only manifest a number of months down the road—all, that is, except for the rapid improvement in blood sugar.

To understand D.J.'s ensuing problem, we will need a brief discussion of glucose, insulin, and the absorption of food here. Purists will find fault with some of my simplifications, but what I am about to put forward is, though simplified, basically true.

When you eat, you chew and add moisture in the form of ingested liquid and saliva. This mash of food then enters the stomach. In the stomach, acid is secreted that helps break down the bonds of the foods you eat. Various other local and systemic chemicals contribute to the movement of food out of the stomach, further digestion, and a sense of satiety. Many of the chemicals localized to the stomach and small intestine have both local gastrointestinal effects and neurologic effects that travel via various nerve pathways to the brain.

As a simplification, once food starts to hit your small intestine, your pancreas gets prepared to secrete the hormone insulin. Insulin has many properties, but its chief one is to get glucose into cells. Glucose is a basic fuel for cellular metabolism. The only cells that do not require insulin to take up glucose are brain cells. When people are obese, they become resistant to the effects of insulin. That is, a given amount of insulin in an obese person is exponentially less effective in doing its job than it is in a lean person. This fact is part of the explanation for the epidemic of adult-onset diabetes in America. Our mean BMI is excessively high in adult life and, consequently, many people develop diabetes, as the insulin they make simply cannot overcome the insulin resistance caused by excess adipose tissue. Indeed, the Centers for Disease Control has, in the last few years, tried to focus our attention on the vast increase in type II diabetes in youth because, with a decrease in exercise and an increase in high-calorie fast foods, many young people are now obese.

In your small intestine, food and other nutrients are absorbed, insulin is secreted into the bloodstream, and (if your caloric needs are matched to your caloric input and your digestive system is in proper working order) you have fuel for your cells and your weight is stable. But this model oversimplifies things, as adults know. There are many determinants to eating, including social pleasures, respite from workaday demands, reward mechanisms in the pleasure centers of the brain, and hereditary body shape and size.

For the morbidly obese, bariatric surgery has proven to be a real benefit. Experience with this surgery is going on twenty years, and the risks and benefits to the patient have become clearer. Benefits include the emotional and physical impact of significant weight loss. And recidivism is not high in contrast to dieting, where regaining weight is more common than not. About 70 percent of people who undergo the surgery do not gain back a lot of weight. And those who start to regain weight can have further noninvasive interventions to help prevent significant weight gain (for example, sometimes a band is used to adjust stomach size). Other benefits are, as expected with weight loss, improvement in symptoms of arthritis, sleep apnea, and diabetes.

However, some serious complications to the surgery merit candid discussion. It was one of these complications that D.J. experienced that was entirely life limiting. In the short term, there can be straightforward surgical complications: leaks at sites of junctions in the surgery, obstructions because of scar tissue, and ulcers that occur around the junctions. Also, the body is not used to the small stomach and having much of the small intestine's length bypassed. Consequently, dumping can occur: Food is not well digested in the new, smaller stomach, and a large amount of it is suddenly sent into the modified and shorter small intestine. This food has osmotic strength (osmosis is the physical process by which particles in a solution exert a force) that draws water from the body into the small intestine. That creates a slurry that now develops some speed and courses through the small intestine, causing cramping and diarrhea. This dumping syndrome frequently occurs within the first three months of the surgery, but sometimes it occurs much later in the postoperative period.

Another rare late complication is the development of new cells that make insulin. It is as if the insulin-producing cells (the beta cells) go on a weight-lifting program—they bulk up. The reasons for this are not well understood, but they likely relate to a genetic disposition in certain individuals and to the enteric hormones (like ghrelin and GLP-1) that circulate locally to enhance digestion and propulsion of food. The name for this proliferation of insulin-producing cells is called nesidioblastosis. The name—coined by a German with a fondness for the Greek language—means "island growth," as the beta cells normally are nested in little islands. In nesidioblastosis, these islands grow larger and even start to grow outside of the ducts of the pancreas, where they are not normally present.

When nesidioblastosis occurs, the usual and customary stimulus to insulin secretion causes a very robust response. The amount of insulin secreted is increased, and it is secreted for longer time, because of all those extra beta cells. One very serious consequence is the development of hypoglycemia. In this situation, there is insufficient glucose available for the brain. In its mildest form, the patient becomes irritable, hungry, and tremulous as the body mobilizes defenses to rapidly get the blood sugar up. But if the hypoglycemia gets worse, the brain is truly starved for glucose, and confusion, coma, and even seizures can ensue.

Nesidioblastosis is not easy to diagnose. It is usually confirmed at surgery to remove the nests of cells. But, as you can imagine, that is a last resort, and many people do not opt for that. D.J.'s symptoms were very severe. When she became hypoglycemic, she would shake, sweat, report that her field of vision had become tunnel-like, and often lapsed into a kind of silly baby talk. Her husband, who had taken early retirement to be with his wife, knew to give her a bolus (a dose) of glucose. But to get glucose when she needed it meant that D.J. had to have an IV in her arm at all times, one that was hooked up to a highly concentrated glucose solution. She had tried other therapies, but D.J. found after eighteen months that the one that worked without fail was the rapid IV glucose infusion.

D.J.'s social situation was terrible because she effectively was trapped in her own home. She was now thin, she had much less joint pain, and her sleep apnea was cured. But she still had some moderate cardiac damage from hypertension, long-standing low oxygen from her sleep apnea, and she could not really function out of her home. Usually she would get hypoglycemic symptoms very soon after eating—no matter what she ate. Occasionally, she would even experience her symptoms after ingesting only water.

D.J. was admitted to the hospital for a recurrence of a small bowel obstruction related to her prior bariatric surgery. One of the residents on my team knew her well, as she had had many admissions over the last few years. All newly admitted patients are seen by senior physicians, and when I sat in D.J.'s room and interviewed her, I was struck by the disability she was experiencing in spite of her otherwise successful surgery. Her examination did not reveal any clues, and when I checked my findings with three prior admissions, her examination was unchanged.

Our hospital team discusses all new admissions in a team meeting. At these meetings, either medical students or interns present their findings. The residents and I try not only to encourage the management of the immediate issues that caused the hospitalization but also to ensure that the team sees the patient as a whole person, not as her condition (in D.J.'s case, a small bowel obstruction status post bariatric surgery with severe recurring hypoglycemia).

When D.J. was presented to the team and I was asked my thoughts, I said, "The best thing we can do for her, once her acute issue of the obstruction is resolved, is to improve the quality of her life. I don't have any answers now, but I want to think on it." We then moved on to discuss the next three patients admitted that afternoon.

I went home that evening and, after spending time with my wife, I retired to my study. D.J.'s situation had been percolating in my mind since I had seen her that afternoon. I used a yellow legal pad and drew a diagram of the many factors influencing her glucose levels, reminding myself of the metabolic pathways and enzymes involved and the design of her surgery. I then built a timeline of her episodes and their relationship to her meals (or lack of meals). Nothing leapt out at me from my yellow pad.

That night, I woke up around 2 A.M. Since college, my brain has often worked in the background on problems I can't solve, and it is not unusual for me to awaken with an insight that has occurred deep in my subconscious. In my sleep it had come to me that nesidioblastosis could occur after obesity surgery—I remembered a case in the *New England Journal of Medicine* from some years before. I also vaguely remembered that there were poisons for beta cells—those insulin-producing cells—called diazoxide and streptozotocin. I thought it likely that D.J. had nesidioblastosis, and I wondered if we could chemically reduce D.J.'s burden of beta cells without any other intervention.

I hadn't thought of either of these medications in years, so I did some research the next day. It appeared that diazoxide was reasonably safe and could be titrated up slowly. As this was not in my usual field, I called a colleague who is an endocrinologist and diabetes expert. He affirmed that it might work, gave me some suggestions as to how to proceed, and outlined what safeguards we needed to have in place for D.J. should this thought experiment turn out to be a bad idea. I discussed it with the team the next afternoon and solicited their ideas. They approved of the concept. I then discussed the idea with D.J. and her husband. I told her that it was an idea, rather than a proven procedure. I told them that I had looked quickly for similar situations in the literature and had not found support for what we were offering, but she was enthusiastically onboard.

We started the medication in low doses, first once a day and then twice a day. Over the next three weeks, we were able to blunt the insulin output of D.J.'s beta cells without completely eliminating their necessary insulin secretion. Her glucose levels remained within normal limits whether she had just eaten or had fasted over twelve hours the night before. The 24/7 IV came out. Now D.J. can go shopping, dining, and even simply take a ride in the car without being tethered to an IV. As she put it to me in a phone call, she is back among the living.

As internists, most of our triumphs are long-term—we get someone to lower their cholesterol, or we manage their hypertension, or we help stabilize them after a heart attack. It is not a daily occurrence that we can tangibly make a difference in someone's life because we have done what we have been trained to do—to think—and by thinking, heal the sick.

THE CPC

MEDICINE IS A PROFESSION AND NOT JUST A
job. That is both the good news and the bad. The public expects,
and the profession demands, that care is not rendered like a
job—where one shows up, does one's work, and leaves without
a further thought about work until the next day.

Part of what shapes medicine as a profession are the
educational activities that have traditionally been utilized to
train doctors. Many of these activities have evolved with the
times, but others remain as cherished modalities for learning. For example, when I was in medical school, for three
afternoons a week for about two-thirds of a year, I spent four
hours dissecting a cadaver. With the advent of CT scanners,
MRIs, and simulators, that time in the dissecting room can
be radically shortened. Indeed, much of what I learned while
laboriously dissecting and standing on my feet can be learned
at home on a computer screen—not all, but a lot.

But some things are not so amenable to self-learning or
home instruction. One such modality is the Clinical Pathologic Conference—the CPC. Many years ago, a patient's diagnosis was often first made by the pathologist, who examined
tissue samples from the patient. Before the advent of noninvasive body imaging, this was often the only way a diagnostic dilemma could be resolved. When CPCs were first introduced as a teaching method in the early twentieth century,

the pathologist often provided a diagnosis from an autopsy and not a simple tissue specimen. As medicine became more advanced in the 1920s and 1930s, often the pathologist could make the diagnosis with a tissue sample of a living patient.

But more than merely providing an answer to a typically complex case, the CPC offers an opportunity for a clinician to show how he reasons about a medical dilemma. This offers the opportunity to see the medical mind at work—raising possibilities, discarding some, keeping a few, and whittling them down to the believed diagnosis. Then the conference is handed over to the pathologist, who has the true and final answer. Clinicians are chosen to do CPCs based on their expertise or on their ability to produce a great show. And, of course, the best CPC discussants are a combination of both.

Academic institutions, as part of their training programs, usually have one or two CPCs (and, therefore, CPC presenters) a month. In addition, the esteemed *New England Journal of Medicine* frequently publishes CPC cases from Massachusetts General Hospital. As I have had the good fortune to work in an academic institution for much of my career, I have attended many CPCs. I look forward to them, both for the challenge of trying to figure out the answer myself and for the wonderful theater they provide. The best CPC presenters not only discuss the details of the case but also use the case as an opportunity to talk about broader academic issues—such as the efficient use of testing, how people frame and think about problems, or the use and misuse of the Internet to drive medical decision making. I've often learned much more than just the answer to a complex and challenging case.

But there is an important aspect of the in-person CPC that should not be ignored: The discussants are truly in a position to either be right or fall flat on their faces. Whenever I have been asked to do a CPC, I start my discussion with the quotation of the gladiators in the Roman Colosseum: "Those who are about to die salute you." This acknowledges an underlying current of the CPC—that young trainees love to see their senior colleagues work but that they nurture the sweet but silent hope that senior faculty members will go down in flames. It has happened to me, and I have seen it happen to others. Part of the drill is to accept the humiliation with grace and good humor.

At our institution, CPCs are held on Wednesdays at noon. The case is not announced to the trainees, but they know there will be food, drink, and a festive atmosphere.

The CPC discussant is given the case approximately ten days in advance and is able to ask a few questions of the preparer, usually a senior or chief resident. However, not too many questions are allowed, and, indeed, the unwritten rule is, "Here's the case, Discussant, deal with it." The cases are all real, that is, they were truly seen in the institution, and none of the details of the case is embellished or changed. The cases are always interesting, they often have a twist, and there is a certain amount of gamesmanship that goes into getting the right answers.

What follows are two of my recent CPC challenges. Each highlights how fascinating medicine can be and how I try to approach traditional teaching venues in a stimulating way for the residents.

CASE 1: A FURNITURE MOVER WITH KIDNEY DISEASE

The protocol of this case was a fifty-five-year-old male who presented to his primary care physician with symptoms of fatigue, nausea, and vomiting. His examination at the primary care physician's office was unremarkable, but lab work showed signs of significant kidney failure. Urinalysis is often very useful in helping determine the cause of the kidney failure, but in this instance, it was useful only in that it was *not* typical of much. Specifically, it had a few cells, a bit of protein, and some clusters of congealed cells we call casts. Had it shown abnormal red blood cells, a lot of white blood cells, lots of protein, or lots of crystals, these findings would have been clues to various types of kidney disease causing kidney failure. But this urinalysis was rather bland and unrevelatory.

The man was married, had two grown children, and worked for years as a furniture mover. He had numerous tattoos, but testing for hepatitis C (which can come from contaminated tattoo needles and can cause kidney failure) was negative. He was tested for HIV (another viral cause of kidney failure in some), and that was also negative. There were no abnormal proteins in his blood, and there was no sign of blood or other malignancy in routine testing. His urine was submitted for a toxicology screen, and it tested weakly positive for cocaine. That one fact, in a lengthy two-page protocol, was the clue. Remember, the patient is unavailable for questioning, and the written protocol is all the information the discussant gets.

The CPC author knew that, two years prior to this discussion, there was a paper published by the Kidney Group at Massachusetts General Hospital about a series (twenty cases) of patients who all had kidney failure and were cocaine users. The cause of the kidney failure was presumed to be a drug called levamisole, which was used to cut the cocaine. Whether it was levamisole or the combination of levamisole and cocaine, in certain people who are susceptible, they developed acute kidney failure. The paper traced the histories of the patients, their pathology on kidney biopsy, and some short-term follow-up.

Fortunately, I remembered reading the paper—though not thoroughly. When I am the attending physician on our wards, we admit and treat many patients with substance abuse problems, and so whenever I am alerted to a paper concerning these drugs, I try to familiarize myself with the highlights. Once I read the CPC protocol, I pulled the article, and, sure enough, this case fit the profile of the other patients.

I didn't want to reveal my hand early, so I presented a discussion of how I approach renal failure—acute versus chronic, involving just the kidney or as a manifestation of a larger systemic disease, and how various laboratory tests can point one toward various likely diagnoses.

I then focused on cocaine and its basic pharmacology. I cited some historical medical figures who were addicted, including Sigmund Freud and William Stewart Halsted, one of the founding four physicians of Johns Hopkins Hospital. I then discussed cocaine's effect on various organs. There was very little literature on its causing direct kidney damage, so I did not mention the kidney when I discussed its effects— the most common effect seen in the inpatient service is heart attack from excess sympathetic nervous system stimulation.

These heart attacks attract the attention of the resident staff because they often occur in young people the same age as the staff. Last, I mentioned the Massachusetts General Hospital paper and how this patient fit the profile.

Once the medical discussant finishes, the pathologist comes to the podium and, in great detail, gives the pathologic findings. The discussant sits in the front row, waiting to be either humiliated or vindicated by the pathologist. On this occasion, I was pleased at the results and with the fact that the pathologist had even gone so far as to discuss his findings with his colleagues across town at Massachusetts General Hospital and had sent the tissue specimen slides to them. All was confirmed as a likely case of levamisole/cocaine-induced acute renal failure. This time, the lion didn't devour me.

CASE 2: TROUBLE ON THE APPALACHIAN TRAIL

The patient was a middle manager, born in India, who immigrated to the United States as a teenager. He presented with a twenty-pound weight loss and chronic gastric upset. He was in a monogamous marriage, was gainfully employed as an engineer and manager, and was normally quite athletic. His particular passion was hiking, which he used to do along the Appalachian Trail every summer vacation. He did not smoke or drink. He had no change in his diet, and he ate meat as well as vegetables. He last visited India some fifteen years prior to this episode of illness and felt fine while there. He had no troubles on his return to the United States. His past medical history was remarkable only for the loss of his right eye five years ago because of an infection.

His examination (according to the protocol I was given) showed a dark-skinned, slender man appearing his stated age of fifty-five. His blood pressure was on the low side (98/64), as was his pulse. He had a glass prosthetic right eye, but the examination of his remaining normal eye was unremarkable. His lungs were clear, and his heart was normal in all respects. There were no enlarged organs in his abdomen. His neurologic examination was normal. Laboratory values showed modest anemia. His serum electrolytes were mildly abnormal, with slightly low serum sodium and slightly elevated serum potassium. A blood urea nitrogen (BUN) test—an indirect measure of his kidney function—was also slightly low. The clues to the protocol were hiking, the loss of the eye, the electrolyte abnormalities, and the low blood pressure.

Our body makes hormones that circulate throughout our entire system and keep our bodily systems functioning properly. These hormones include, to name a few, thyroid hormone, insulin, and cortisone. If the cortisone level is low, our body's ability to handle sodium, potassium, and water balance is interrupted. Furthermore, the most common clinical symptom of cortisol deficiency is gastrointestinal upset. So it was apparent to me that cortisol deficiency was a significant issue in the case: But how to tie in the significance of the eye enucleation (eye removal) and hiking the Appalachian Trail?

Well, in the caves of Georgia, where the patient hiked the Appalachian Trial, there is a fungus called histoplasmosis. When it disseminates in our body, histoplasmosis can affect the eye and the adrenal glands, the source of the cortisol hormone. Other indolent infections, like systemic tuberculosis, could do the same thing, but only histoplasmosis fit with the hiking history.

My working hypothesis was that this man had camped out while hiking in caves. He had been exposed in those caves (most likely through bat guano) to histoplasmosis. The histoplasmosis had over time, and quietly, disseminated to his eye, requiring it to be removed, and had likewise systemically spread to his adrenal glands, where, over time, it caused adrenal insufficiency.

Once I made up my mind about the diagnosis, I decided to look into the physician who had first described adrenal failure caused by destruction of the adrenal glands. That fellow was Thomas Addison of Guy's Hospital in London in the 1830s, and I was able to give a brief talk about his discovery of the condition that now bears his name (Addison's disease is adrenal insufficiency from destruction of the adrenal glands). After discussing Addison, his life, and his history, I laid out my thesis for the cause of the patient's symptoms and findings. And I was gratified to learn that I was correct when the pathologist took the lectern.

I hope you can see the point of the clinical pathologic conference. They are always fun cases, often unusual, and, like a difficult puns-and-anagrams crossword puzzle, the clues are there. It is easy to get the wrong answer, and, as the exercise is a platform for the discussant, the case can be used to highlight how a clinician diagnoses conditions. But the case can also provide an opportunity to discuss history, medicine, the latest literature, or anything else of value to the audience, who comes to be both entertained and enlightened. Many educational traditions have changed in medicine, and many are amenable to learning at one's own pace. But some things can only best be transmitted by watching an experienced doctor grapple with a problem publicly and with some risk in the atmosphere. I hope they never disappear from the educational scene because, like rewarding theater, the best CPCs leave the audience with a suspension of disbelief and enhance their humanity.

LET THE FACTS SPEAK FOR THEMSELVES

"DR. MUSHLIN, WILL YOU COME IN HERE, PLEASE?"

Nicole was a nurse on the tenth floor I had known for eight years. We had worked often together. She was resourceful, independent, and very capable. For her to ask me to help her was beyond unusual.

I entered the room of Ms. B., one of our new admissions for the day on the Intensive Teaching Unit. Ms. B. was screaming at Nicole and calling her all kinds of names, none of them repeatable. I asked the patient if I could please approach her bedside, and she said, calmly, "Certainly." I asked Ms. B. what seemed to be the problem, and she said that Nicole, like most nurses she had encountered, simply didn't get it and that Nicole "didn't understand the kind of pain I'm in."

I told the patient that I'd be happy to talk with Nicole about her observations regarding Ms. B., but that the verbal abuse of Nicole would have to stop right now, and that I would return to interview and examine her. The patient readily acceded to this, and Nicole and I left the room.

Once outside, Nicole told me that the patient had been extremely uncooperative. When they have a new admission, nurses have to assess the patient, go through a checklist of patient needs and issues, examine the patient, and, after all that, complete a lot of paperwork to document everything. Nicole was unable to get started with the patient, as she refused to participate in any of the customary questions, and rapidly started to call Nicole all sorts of insulting names.

Nicole had already reached out to her nurse supervisor on the shift. I told Nicole that, if it was okay, I'd go back in. I wanted to get more information from the patient that I could share with the nursing staff, and I also wanted to try to get the patient to behave in a more appropriate manner with all of the staff. With the permission of my nursing colleagues, I returned to Ms. B.'s room.

Mrs. B. was a fifty-year-old Caucasian female, and she was pacing in her room. I could not get her to stop pacing, and although she was not a threatening presence, I didn't feel I could predict what her next move would be. She refused to tell me where she lived, where she was born, or whether she was married or single. She did know some of her medications. She knew that she had had gastric bypass surgery approximately five years ago "somewhere in Connecticut" and that she had had a number of admissions for small bowel obstructions after that "botched" surgery. She had a placement of a tube in her jejunum (part of the small intestine) because of her recurrent small bowel obstructions, and that was now the only way she was getting any nutrition. However, the area around where the jejunum tube was inserted was infected again and so couldn't be used. She had been admitted to our hospital just ten days earlier, on the nonteaching service (the nonteaching service is staffed by hospitalists, those physicians who work exclusively in hospitals; the teaching service has staff physicians who have trainees with them), and she had left the hospital against medical advice. When I asked her why she had left, she stated, "They didn't understand me, and they took really bad care of me." She wanted me to know that she had a lot of pain: in her abdomen, in both legs, and in her back. This pain was not new, but it had not been properly addressed for years, and it was getting to be a "pain in the a** to keep telling people how sick I really am, and them ignoring me."

I knew that I was not going to get much more information from her at this initial encounter, but I was heartened to hear that she had recently been to our hospital—the recent record might fill in some of the blanks in her history. And though she was not outright hostile or rude to me, it was clear she found me a nuisance and was putting up with my questions only because I was a means to the end she wanted.

We keep electronic records, so her previous admission was easy to find; she had stated that her jejunum tube was not working properly and that the site around it was painful. She had been admitted to the medicine service, and our surgical colleagues had been consulted. They had found the tube to be intact and well positioned in her small intestine. They had thought the surrounding area on her skin was not infected. Soon thereafter, the patient had signed out against medical advice, and—reading between the guarded lines in the discharge summary—it appears that there were a lot of conflicts between the patient and staff, especially about the timeliness of the staff's response to her requests (or, more likely, demands) and the patient's insistence on strong doses of narcotics. Apparently, the patient became very agitated, frustrated, and left. The doctor on duty had well documented that she had spent time trying to persuade the patient to stay, but to no avail.

As concerning as that discharge was, even more concerning was, once again, the lack of information about Mrs. B.'s past. There were no data as to where she was living, where her prior surgeries were done, or whether she was single or married; essentially, there was no past history or social history available. This is most unusual. Physicians, indeed all treating staff, need to know where a patient is coming from, literally and figuratively—where do they live (or are they homeless), who can serve as a caregiver (or are they truly alone in the world), and where were these operations done (in case we need information from the hospital about a prior procedure now gone awry)? These questions are standard and essential. And the lack of data in this area was a red flag to me. But house staff can be busy, and if they think an admission is a soft admission (that is, someone who really does not need to be admitted), they often do not press for standard information; they rather address the problem that precipitated the admission and move on.

So, I decided to see what I could find out for myself. I also thought it was best to let the nurses do their admission procedures and let the intern and resident assess the patient. I would go see the other admissions we had and then return to the patient's bedside and try to fill in the blanks.

When I returned to the floor a few hours later, Nicole just rolled her eyes at me and said that this patient was one of the most difficult she had ever had to care for. Nicole related to me that everything was a negotiation. The patient did not want the pillow issued to her, and she did not want any of the medications that Nicole brought in and was supposed to observe her taking. The patient said that the recently started

IV was burning and that she needed a new one. The patient refused to give a urine specimen as ordered by the intern. It went on and on in this manner. In addition, Nicole said the patient was very rude and offensive and called her all sorts of names. And, as if that were not enough, Nicole suggested that I simply wait out by the nurses' station and observe. So I did. A few minutes later, Ms. B. came out and stood in the hallway. She addressed everyone who came by, telling them, unbidden, what a lousy nurse she had, how this hospital was terrible, and how she had no intention of having them murder her.

Clearly, this patient was going to be a challenge to manage and to try to help.

I told Nicole that I was going to try to get some more information from Ms. B. and try, with my advanced age and gray hair, to see if I could get her to cooperate with the treatment plan that the younger doctor had outlined. Ultimately, I was going to see if any of us on the team could form a therapeutic relationship with her to help care for her.

Just before I meant to enter her room, the personal care assistant who was taking her vital signs told me that Mrs. B. had a fever of 104°F. A temperature of 104 is high, and it requires an immediate response. The intern, resident, and Nicole went in to administer to the patient. I felt very confident in the second-year resident, so I decided that I would give them all plenty of space, and I did not go into the patient's room. Again, I went to see a fresh admission. When I returned, I listened to the team's assessment and concurred with the institution of broad-spectrum antibiotics, culture of the jejunum tube skin-entry site, chest X-ray, urinalysis (assuming Mrs. B. would agree to give one), blood cultures, and a complete blood count. I asked to be kept informed and discussed using a cooling blanket if her fever increased. I also suggested that we alert the third-year resident in charge of the medical intensive care unit that we had a suddenly very sick patient who might need a transfer to the MICU if things did not go well. And indeed, things did not go well.

Mrs. B.'s white blood count was highly elevated, and the distribution of cells implied an acute infection. Her blood pressure started to drop and did not respond to vigorous fluid resuscitation (this is often a sign of gram-negative bacteria in the blood), and an arterial blood gas analysis showed the patient had rapidly developed acidosis, a condition that is a sign of significant biologic decompensation.

By early evening, all was arranged for her to be transferred to the MICU. I spoke with my colleague who was the MICU attending about her, and I was able to go home. But I was troubled: troubled by her behavior to Nicole, troubled by her acute decompensation, troubled by the fact that, on reflection, maybe I hadn't interviewed and examined her already because I thought she would be difficult, and that made me somewhat remorseful. And, finally, I was troubled that none of the team realized that she might have been very sick when she was admitted.

The next morning I stopped by the MICU on my way to the floor. I was pleased to hear that the second-year resident had already been there to check on her. Mrs. B. was afebrile, her blood pressure was normal, her kidney function was improving, and she no longer had acidosis. The MICU team had done a great job throughout the night. The cause of her acute decompensation was still a mystery, but, as often happens in these situations, matters became clear in a day or so. Indeed, by the late afternoon, our team received a page from the hospital's microbiology lab informing us that all of the patient's blood cultures were positive for not just one bacterial species (as is most typical) but three. They could not characterize the bacteria at this stage any further, but that is common, and at least we had a path to follow in finding out what had happened to her. Usually when there are multiple bugs in the bloodstream, it implies a source from the gastrointestinal tract. We called the MICU team and told them of the early findings. They informed us that it was likely that she could be transferred back to our team by the late afternoon.

When she returned to the floor, she was too weak to give the nursing staff a lot of trouble, and when I went to see her, she was passive, but she didn't answer many of my questions. I inferred she had been living in other hospitals or rehab facilities in Connecticut and New Hampshire for the last two years; she had no immediate family; she had been using high doses of narcotics for vaguely defined pain for over four years; and she had lost fifty pounds over the last three years. It seemed she didn't like to try to eat, as she felt it provoked repeated small bowel obstructions, and she was dependent on her jejunum tube for feedings. But, as it had been repeatedly blocked or malpositioned, she had not often been fed through it.

Her examination did not reveal any obvious source of her bacteremia (the presence of bacteria in her blood). She had loose skin folds, typical of people who have lost a lot of weight, and she had numerous scars on her body over venous access sites, implying many surgical attempts to place IVs.

In the MICU, a central line was placed in her internal jugular vein, and she stated how pleased she was to have it, as now her medications could be administered without discomfort (her peripheral IV from the day before, she stated, burned all the time). Her blood pressure was good, her lab values were all improving, and we got the results of her polymicrobial bloodstream infection. The bugs that were growing were ones most typically found in stool.

The infectious disease consulting team helped narrow her antibiotic choices and recommended that she have two weeks of intravenous antibiotic treatment. When we discussed with them their thoughts as to the origin of her bloodstream infection, they thought it most likely that she had self-injected feces into her bloodstream or skin.

Their strong suspicion, and the increasing red flags in her history and exam, made us strongly consider that we were dealing with a patient with Factitious Disorder, also known as Munchausen's Syndrome—a very serious psychiatric disorder in which a patient fakes symptoms, distorts illness or treatment progression, induces injury to himself, and even persuades unwitting physicians to offer care that is not indicated.

A colleague of mine shared a Factitious Disorder patient experience in which the patient had somehow persuaded numerous physicians to sequentially amputate four fingers of her right hand. I never got the details from my colleague, but more than likely, he never got all the facts, either, as that is typical of these patients. They hide things, they often cause their own illnesses, they are often addicted to narcotics, and they often spend years in hospitals. And when they are confronted with their illness, they often decamp rapidly, go to another state, present to a fresh hospital, and start the cycle all over.

This is a very serious condition. These patients run all the risks of hospital-acquired infections, all the risks of multiple surgeries, and all the risks of multiple invasions to their bodies. Many die. I do not pretend to understand the underlying psychodynamics, but Mrs. B. was the third patient I've seen with Factitious Disorder in my long career.

Factitious Disorder is a diagnosis of exclusion, but I felt we had enough data to go on. Our team also had an ethical dilemma. Mrs. B. had not formed a therapeutic alliance with any of us. She withheld information, and she had almost certainly harmed herself by self-injecting her feces in an effort to get continuing care. The fact that she had self-injected once, and now had direct intravenous access by means of the internal jugular vein, meant she could certainly do this again. Our active concern was how to firmly confront her with our concerns yet encourage her to let us finish her course of treatment. We consulted our psychiatry department, as well as our ethics team. We had two separate meetings with all of them over the next three days. We all felt that when Mrs. B. was confronted, she would sign out against medical advice and just go elsewhere. We debated how best to therapeutically tell her our diagnostic thoughts and how best to try to enlist her in her own care. And last, we decided to have a treatment plan in place if she did suddenly leave and then returned to our emergency department in shock or with high fever. Normally, when we have an ethics consultation, the ethics team interviews the patient and solicits his or her thoughts about what the issues are. This was the first time in my experience in which the ethics team thought it would be counterproductive to even meet with Mrs. B. in advance.

It was determined that I would discuss our team's thoughts with the patient and that the intern (who had primary responsibility for the patient) and the resident would also be in the room. I didn't know how the conversation would go, but I wanted it to be a learning experience for all of the physicians on the team.

As predicted, it did not go well at all. The patient absolutely denied that she had caused her bacteremia. She personally attacked the intern, stating that the intern "had it in for me from the beginning," and she personally attacked me, hoping "that you burn in the eternal flames of hell forever." She got up, went into the little wardrobe where her clothes were stored, and refused to sign the form we offer patients who leave against medical advice.

These unfortunate patients present many problems in our current hospital system. Rarely is the same senior physician responsible for their care on their repeated admissions. The patients are adept at manipulating staff to get what they want. Often they present to the hospital with serious illness, mostly brought about by their own hand. Furthermore, what is the ethical line to refuse care to a patient—when can she be denied admission to the hospital? Is it appropriate to repeatedly admit a patient who refuses to cooperate with her care team? And, in general, these patients take up a tremendous amount of staff time—to the detriment of other patients on the floor.

What our team did (and I still don't know if this was the right and best approach) was to consult with the hospital lawyers about setting parameters for this patient should she return. We also put a care plan in her record that would be implemented in the emergency department: If she was not gravely ill, she would have to consent to having her narcotics tapered during the admission, and boundaries were to be established on what was unacceptable behavior to the staff. Should she violate those boundaries, she would be escorted out of the hospital.

To date, she has not returned. My hope is she hasn't managed to kill herself with her self-harming behaviors. But I strongly suspect she is currently hospitalized in some other state, harming herself, berating the staff, and stealing time and energy from the other patients on her floor.

COUGH

DR. C. IS TALL AND VERY LEAN, AND WHEN I see him, I always think that he is one of my very true liberal friends. He is on the side of the working man, the disenfranchised, and the dispossessed. He tilts at windmills, daily, and never tires of it. And he gets into trouble for it. An administrator took him to task when he bought medications for one of his clinic patients who couldn't afford them. To some, that's a boundary violation; to me, that is simply Dr. C. His wife is just as skinny, but I rarely see her, as she is running an HIV prevention project in southern Africa. They are apart over six months out of the year. I have had the honor of caring for many physicians in my practice. Most were high-achieving, highly motivated people. But no physician that I cared for had more intrinsic integrity and respect for mankind than Dr. C.

Doctors as patients fall into a few categories. The first is the know-it-all: "I, myself, can take care of what ails me." These physicians often come to me very late in the course of their problem. And when they do come in, they are sheepish about their self-medication or self-ordering of tests. The next category is the hypochondriac. A new gray hair is cause for a search for heavy metal intoxication. Not being able to recall the year a paper was published that a physician wishes to cite is cause for a full-blown workup for Alzheimer's. These physicians often acknowledge their hypochondria but are helpless to stop it. My job in our mutual relationship is to play the traffic cop—"No, you can't have yet another MRI for this symptom." The third category is the denier. Even in the face of the most striking symptoms or physical findings, these physicians prefer to think that nothing is wrong. By the time they come to the office, it is often evident to me, without any testing, what the problem is—but getting these physicians to treatment proves to be the real challenge.

Dr. C. fell into the third category. He could have an arrow piercing his chest, through and through, and he would deny that it was troublesome or abnormal. So when he came to see me, my alertness level was on overdrive. Dr. C.'s wife had insisted he come in and see me. She had noticed a persistent cough, and she thought, as she had been in Africa for a while and had just returned, that Dr. C. had lost significant weight. Fever and cough are a bad combination, and usually something is going on—often in the infectious disease domain. Doctors are exposed to many odd infections and often find out only after the exposure what they were at risk for.

Dr. C. conceded that he had lost weight. His tiny waist was even tinier. He also was certain he had had a fever for at least a month. He would feel warm in the hospital every afternoon. A few times he took his temperature when he got back to his apartment, and it was always over 101.5°F. He denied chills. His appetite really was no different. When he ate, he had no pain in his throat or in his abdomen. He had no change in his bowel movements. He denied a rapid pulse or emotional lability. His cough was present both night and day, and occasionally he would awaken in the middle of the night and realize he was coughing. He would raise some frothy white sputum. There was no blood in his sputum. He had no wheezes and no remote history of asthma. He did notice he readily got short of breath. Climbing a small hill near the hospital provoked a cough and shortness of breath. He had no rashes, no joint pains, or swellings. He did not have any pain or change in the shape of his fingernails.

When I examined him in my office, in the late afternoon, he looked exceedingly thin and haggard. He had a fever of 102.4°F. His pulse was slightly elevated at 88. His respiratory rate at rest was 16 and his oxygenation at rest was 92 percent (normal is around 97 percent or higher).

His lung examination was not very impressive. He had normal lung movement and a few coarse scattered bubble-like noises we call rales in the right lower lobe area and the left middle lobe area. His cardiac examination was normal, and he had no enlargement of the liver or spleen. I ordered a chest X-ray and lab work immediately and told Dr. C. I would look at both before I left the office. We arranged to speak by phone later in the evening.

Dr. C.'s chest X-ray showed diffuse and scattered infiltrates, which, if the cough were new and acute, I would call typical of pneumonia. But with the diffuse and scattered nature of it, and with his cough and weight loss having gone on for a while, I doubted very much that this was a simple pneumonia. His white blood count was modestly elevated at 15,000, and he had a slight excess of lymphocytes in the differential. His platelets were also slightly elevated at 550,000, and he had moderate anemia, with a hematocrit (the ratio of red blood cells to blood volume) of 33. He did not have an excess of a type of white blood cell called an eosinophil. A marker of systemic inflammation, hs-CRP, was markedly elevated.

Liver tests, kidney tests, and thyroid tests were normal, as were basic screening tests for rheumatic diseases such as systemic lupus, rheumatoid arthritis, polymyositis, and granulomatous angiitis. I also ran a test for sprue (celiac disease), as I had seen a case of sprue (which can cause weight loss) with associated pulmonary disease—but the sprue test was also negative. In fairness, some of these tests took a few days to return, but I was able to communicate the basics to Dr. C. that evening.

I called Dr. C. and told him he had lung involvement diffusely, signs of inflammation in his blood work, and no sign of malignancy as the cause of his cough and weight loss. Of course, like most faculty, Dr. C. had accessed his own record ahead of me, so we could move right into the discussion of what was next.

I told him he needed to see a pulmonologist, and soon. I told him that, in my limited experience, he likely had a variant of bronchiolitis—an inflammatory condition where the tiny air sacs (alveoli) get inflamed and full of fluid. There are a number of types of bronchiolitis, but they tend to be managed the same way—by using high-dose steroids to cool off the inflammation. I also told Dr. C. that I had seen a condition of bleeding into the alveoli look like this on chest films, but he was not coughing up any blood, nor did he have any bleeding abnormalities on his screening lab tests. And last, I mentioned that this could be a type of lung cancer—something the patient was already fixated on.

Dr. C. was onboard with seeing a specialist, so I e-mailed a colleague who I had known since I was an intern and he was a third-year medical student. My friend and colleague was one of the best, if not the best, pulmonary experts in Boston, and I knew he would make the diagnosis while also being sensitive to Dr. C.'s personality and sense of denial.

Doctors often get to go to the head of the line, especially in academic medical centers, and Dr. C. was seen the next afternoon. Diagnostic testing was planned, which included pulmonary function tests, a chest CT scan, and a bronchoscopy. The latter is a procedure in which a tube with a fiber optic light source is passed into the lungs. The pulmonologist can direct the tube to the area seen on the X-rays where the inflammation resides. Once the tube is at the correct location, the pulmonologist removes some of the fluid there, washes it, and then analyzes it. Often the pulmonologist will do a biopsy as well and have the pathologist examine that lung tissue.

Dr. C.'s chest CT scan, unsurprisingly, showed more extensive involvement, diffusely, than the standard chest X-ray. His pulmonary function tests revealed nothing unexpected. However, Dr. C. was frightened about the invasiveness of the proposed bronchoscopy. He was very insistent that it not be done. He wanted a course of antibiotics to see if he had a simple community-acquired pneumonia.

My colleague the pulmonologist is one of the kindest and gentlest men I know. He was flummoxed by Dr. C.'s intransigence, as he felt that Dr. C. was placing himself in danger, allowing his inflammatory process to gather further steam while waiting for the likely ineffective antibiotics to play out their course.

I, too, could not persuade Dr. C. to proceed with deliberate speed to a bronchoscopy. So we prescribed two antibiotics, and Dr. C. agreed to follow up every ten days for the next three weeks. His fevers, his lack of appetite, and weight loss continued. His cough continued. At the end of two weeks of the three-week course, with the persuasion of his wife, he threw in the towel and agreed to a bronchoscopy.

The bronchoscopy and washings showed just inflammation and no concern for hemorrhaging into his alveoli. The biopsy confirmed the pulmonologist's leading diagnosis of cryptogenic organizing pneumonia (COP), which is one of the many causes of alveolar and distal bronchial inflammation. Its cause is unknown—but, as in so much of medicine, the patient is simply unlucky. It is seen most often in patients in middle age. The pathologic findings under the microscope are quite specific and lead to the diagnosis. And, fortunately, COP tends to be responsive to steroid treatment.

Steroids are a double-edged sword. They are potent anti-inflammatory medicines and can cure or abort many medical conditions. But in high doses, as were needed in Dr. C.'s situation, they have risks. They can cause bone fractures (especially in the hips), emotional lability, sleep disturbances, hypertension, drug-induced diabetes, and susceptibility to infection, as well as other conditions. So their use is not trivial, and physicians always discuss the risk/reward profile with the patient.

By now, Dr. C. was frightened and eager to start treatment, and he responded well to it. His cough improved within ten days, he started to gain weight, and successive CT scans of his chest showed slow and continued improvement.

His fevers and sweats all vanished after the first week of treatment.

He suffered no ill consequences of the treatment, and within six months, he was off steroids completely. Follow-up CT scans and examinations, both by me and my pulmonary colleague, showed no recurrence. With continued good fortune, his illness will not recur.

We never had any inkling as to why this happened to Dr. C. In all respects, he was otherwise healthy. He had no significant allergies, was never a smoker, and had no occupational exposures. He was late in coming in to be diagnosed because of his personality and because he had an attitude about his care so common in physicians—denial and a sense that he should manage his own illness. But fortunately, that didn't seem to hurt his response to treatment.

These days, I bump into Dr. C. quite frequently in the corridors of the hospital, and we rarely talk about how he's feeling; we mostly talk about the music he and his wife are listening to lately.

GREAT IMITATORS, PART 1

ONE OF THE REASONS I WAS ATTRACTED TO internal medicine as a medical student was the adventure of figuring out how to diagnose the great imitator diseases. At conferences, an esteemed clinician would often be given a particularly opaque case for diagnosis by a smug chief resident (who knew what the correct diagnosis was), and the clinician would test the case against the great imitators, those diseases whose manifestations are so protean and difficult to parse that they have to be mentioned in any complex situation.

As a young clinical student, I was fascinated by how these experienced clinicians could raise, and discard, these great imitators, focusing in each case on what was characteristic of the illness and how the patient's condition just didn't quite fit. I hoped that, one day, I would have enough experience to know why, for example, a case could not be sarcoidosis but instead was more typical of amyloidosis, and why, unless we had more information, we could not exclude systemic lupus or, even possibly, systemic mastocytosis—you get what I mean. Since those long-ago days, I have always hoped that a great imitator illness would walk into my office, and I hoped, humbly, that I would have the knowledge and perseverance to figure out what it was. Here is one instance.

Mr. K. was reasonably healthy. I had inherited him from a colleague who had recently retired. The patient had seen a couple of doctors prior to visiting me, and essentially he was interviewing me to see if we were a good fit. This is something I have never minded, as I feel it is always a productive use of our mutual time to find out something about each other's style. And should a patient decide that he and I are not a good fit, it is better to find out early, before some serious problem is before the both of us, with the patient clearly having no confidence in my abilities or persona.

Mr. K. was a middle-aged Vietnam War veteran who, after the war, had become a criminal lawyer in Boston. He was quite a character—his worldview was very cynical, and it was colored by the industrial quantities of vodka with which he fueled his days. Aside from his excessive drinking, he was otherwise healthy, moderately overweight, a nonexerciser, but no other red flags. He was happily married with three good kids, one of whom, he was proud to tell me, was a freshman at Harvard College. He knew I found him to be a character, and sensing he had an interested audience, he entertained me in this first visit with stories of criminals, bent judges and legislators, and the best places in town to get chowder.

His exam was unexceptional but for a slightly enlarged liver. I counseled him about drinking, asked about guns at home (none), and drew some labs. The labs were not bad, but his liver tests were a little elevated. When the liver is being pickled with booze, usually two hepatocellular (liver) enzymes are elevated, and usually in a certain ratio. My new lawyer friend had mild elevations of these enzymes, but in a

1:1 ratio, and what was truly elevated was a different enzyme called an alkaline phosphatase.

As in so many situations in my world, when you start turning over rocks, you find even more rocks that have to be turned over. I called the patient a few nights later, told him about the liver tests, and used those results as a platform to say that he really should cut back on the drinking. But I also shared that the profile of the liver abnormalities was a bit unusual, and I wanted to repeat his tests in around three weeks. He was amenable: He agreed to try to cut back on his vodka martinis, and he agreed to have the repeat labs. And when he did come back for the repeat testing, all of his liver tests were normal.

Happily, I told him to keep up the good work and thought nothing more about it. I saw him again around eighteen months later. He was in for a general checkup, and he told me he had continued to drink in moderation, as both his wife and I had "put the fear of God in him" about his liver. So it was a bit of a surprise when, on repeat liver testing this go-round, his liver tests were quite a bit more abnormal and, again, the alkaline phosphatase test was much higher than the hepatocellular enzymes. Feeling guilty, and worried that the lab test results eighteen months earlier were a lab error or mix-up, I pulled out a lot of stops. I ordered an ultrasound of the liver, which did not show masses or a blocked or inflamed gall bladder and, indeed, looked quite normal. I ordered a lot of liver tests: screens for liver cancer and viral hepatitis (in all its forms), screens for liver inflammations like primary biliary cirrhosis and iron overload. I also checked for a copper transport disease that affects the liver, Wilson's disease.

All negative. And repeat liver tests after this big workup were negative once again. Now this cynical criminal lawyer was looking at me, thinking that I was some sort of idiot—I had ordered a lot of blood testing, now several times, and ultrasound studies. He felt fine, and I felt stupid. It was clear my patient was not too impressed with my diagnostic acumen.

I try to keep my ego in check when I deal with patients, so in spite of my embarrassment, I tried to feel what the patient across from me was feeling, and, as has so often happened in the past, it helped me find a way to be helpful. I told Mr. K. that I would like to run his case by an esteemed and very senior colleague who had written the *Textbook of Gastroenterology* some years ago. I said that I would like to run his situation by my colleague, and perhaps I could ask my colleague to see him. I could see the patient knew I was trying, and he said, "No more doctor visits, no more blood drawing. Just talk to the guy and let me know."

I did talk to my colleague (I bought him lunch in our cafeteria two days later) and told him all I knew. He was also at a loss. He graciously offered to see the patient, but I told him that, currently, that was a nonstarter. We agreed to stay in touch, and my colleague reassured me that, in his experience, an answer would always eventually declare itself.

A few years went by, and I continued to see Mr. K. as a patient. Most visits were for annual checkups, but occasionally he had the odd cough or cold. He continued to drink in moderation. By then he had put two of his three children through college, and his eldest, the apple of his eye, was at Boston College Law School.

Then one day he came in with a rash. Most internists are only fair dermatologists. Common things, like allergic reactions to poison ivy, are easy. So, too, in our area, is Lyme disease. Psoriasis is very common and easy to diagnose. So are the simple skin cancers. But my patient had a few raised reddish lesions on his left arm and a similar, but larger, one on his left leg. The margins of the lesions were a bit firm, and I got the sense that the lesions extended to his subdermis (the lower layer of the skin).

I was puzzled (once again) and sent Mr. K. to a dermatologist for a diagnosis, and I warned him that the dermatologist might want to do a biopsy. He was amenable, but when he appeared in her office eight days later, the lesions were almost completely gone. There was just a little discoloration to the skin, but all the firmness was gone. The dermatologist thought a biopsy would not be productive and gave the patient an emollient to speed healing.

I do my best thinking in the early morning, when it is quiet and my mind is not cluttered with a million things. And, truth be told, I probably do my very best thinking during my morning shower. Often the thinking that is going on is in the deepest part of my brain, and I am unaware of my insight until, like the Loch Ness Monster, it arises from the depths.

I called Mr. K. that day and said I wanted to talk about his service in Vietnam. He started to tell me what life was like as a marine and the way people treated him when he came back (which was poorly). I stopped him midsentence and asked, "Did you ever have leave?" "Sure, Doc," he replied. "We had R & R [rest and recreation] in Subic Bay in the Philippines, and also in Perth." "Did you ever get the clap?" I asked. He paused, then said, "Doc, everybody got the clap." I asked if he had been treated for it, and he said, yes, he got one shot of penicillin. I asked if he had been ever diagnosed with syphilis, and he said, no, not to his knowledge. I told him that the fleeting liver tests, and now the fleeting skin lesions, could represent very old syphilis. The only way to tell was to do some further blood tests.

He wasn't thrilled about the prospect, and he was worried that, if it were true, he had harmed his wife. But I told him that, if he did have syphilis, the causative agent—*Treponema pallidum*—was contained within his body, and he was not a risk to her. He came in for tests. The initial screening test we did, called a VDRL test, was positive with a low titer. A very specific follow-up test for the *Treponema*, called an FTA-ABS test, was also positive.

My patient was at the mercy of one of medicine's great imitators, syphilis. Almost certainly it was acquired during his recreational activities when he was a marine. It went undiagnosed and remained mostly dormant in his body, except when it reemerged and caused lesions that were fleeting and, until now, unexplained.

When syphilis is untreated, the body may heal itself completely. Or, in around 20 percent of cases, the disease can go on the down-low and hang around the body but not cause much harm for some years. When it does cause harm, it does so in one of two (usually mutually exclusive) ways: It can cause fleeting groups of oval granulomas (masses of tissues) in various organs. Syphilitic granulomas are called gummas, and it was the gummas that appeared on his skin that enabled me to think of the diagnosis for his condition. Undoubtedly, but unproven, the fleeting gummas in his liver caused the intermittent elevations of his liver function tests. The gummas do contain active treponema spirochetes.

The other way late latent, or tertiary, syphilis (as it is called) manifests itself clinically is in the central nervous system. There it causes two large categories of disease (again, usually one or the other, but not both). One is called general paresis of the insane, in which executive function is compromised: The patient may be without the usual checks in personality and may act bizarrely, be depressed, or become psychotic. It is said that Winston Churchill's father, Sir Randolph Churchill, had general paresis and would pound on the desks in the House of Lords, screeching like a monkey. The other central nervous system manifestation is in the posterior columns of the spinal cord. By damaging these, and the ganglia leading to them, late latent syphilis severely impairs a patient's ability to sense position and maintain balance. Indeed, in the Prussian Army in the late nineteenth century, one test of the soldiers for late latent syphilis was to have them stand at attention and close their eyes—if they fell on their face, they were deemed to have late latent syphilis.

But late latent syphilis can have other effects on the body, which is one of the reasons it is called a great imitator. Back in the beginning of the twentieth century, the great Canadian physician Sir William Osler stated that "to know syphilis is to know all of medicine." Because the disease was very common in Western Europe, much intellectual effort was made, and financial expenditure given, to find an antidote or curative therapy. The heavy metals, such as gold and arsenic, showed some promise, and Salvarsan, a therapeutic arsenic compound, was used with modest success. But it was the serendipitous discovery by Alexander Fleming of penicillin, extracted from the penicillium mold, that proved to be curative. To this day, the treponeme is quite sensitive to low-dose penicillin if administered over a few weeks. This is the treatment that Mr. K did *not* get. What he got was a blast of penicillin, just once, and that treatment was inadequate. Penicillin leaves the body quickly unless it is bound to a slow-release platform. That onetime penicillin shot did not cure Mr. K.'s disease, which instead went quiet for years, only to recur as a diagnostic puzzle. He endured a lumbar puncture for more lab tests to see if there was evidence of syphilis in his central nervous system. As he had a low titer there but no manifestations, my infectious disease colleagues recommended he have a daily three-week IV infusion of moderately high-dose penicillin. It seems to have worked. Mr. K. has had no recurrence of any symptoms, and he invited me to his son's law school graduation, which I was happy to attend.

GREAT IMITATORS, PART 2

TUBERCULOSIS PLAYED A LARGE PART IN MY life, and it undoubtedly contributed to my becoming a physician. I grew up surrounded by its effects, and, as is characteristic of the illness, it continues to lurk all over the world, killing the impoverished, the imprisoned, and the unwell, as though antibiotics to treat it never existed.

My father was a first-generation American. His parents were immigrants, and they were poor. Once in the United States, they lived in close quarters in tenements in New York City. My father had three siblings, and his next-eldest brother came down with tuberculosis in adolescence. Soon, thereafter, so did my father. Their exposure was probably from a frail uncle who had lived with them in their early years. Extended families living in very close quarters were the norm, making them a fertile breeding ground for the spread of tuberculosis.

Both of these young men came down with the disease during the Great Depression, when the family had no money. Worse, having a family member with tuberculosis (or, as it was generally called, consumption) stigmatized a family, so the family took pains to hide it. Although doctors were expensive, they were consulted. The older brother was deemed to have a severe case of the disease, and he was sent to a public tuberculosis sanatorium in Upstate New York. In the early twentieth century, until the late 1950s, most states had such sanatoria. They were there not only to treat patients but also to isolate them from the general population, and as inmates of state sanatoria, patients became wards of the state. Soon after his older brother was shipped off to the sanatorium, my father became more ill and joined him. Within eighteen months, his older brother was dead. They were close, but there was no survivor guilt; my father was focused on staying alive. As he was a ward of the state, he learned how to make peace with institutions: how to not struggle against a system, how to perceive the larger system around him, and how to focus on life when he was surrounded by people who were dying. On his own initiative, he got a scholarship to Trudeau Sanatorium in Saranac Lake, New York, a private sanatorium with a good reputation for its cure rate and its modern attitude. Many well-to-do patients were there, as well as many physicians-in-training, as young interns and residents were at high risk for coming down with tuberculosis in those days. My father had to do work-study while there (helping to collate patient charts and tabulate patient statistics), and he was housed in a small cottage with three other roommates. Those in the cottages on the grounds of the sanatorium who survived often became lifelong friends.

There were no antibiotics for the treatment of tuberculosis then. Sulfa compounds were just coming into wide use, but sulfa is useless against the bacteria that cause tuberculosis. The mainstay of treatment was rest and exposure to cold air. I have pictures of my father smoking on the porch of his cottage in a bulky raccoon coat in the height of the Adirondack winter. In his case, the standard treatment was not successful, and his illness worsened. So he was treated by resting his most affected lung. This was done by taking him to a room—imagine an old 1930s movie image of a white enameled hospital outpatient room—where a large syringe was inserted into his thorax and air was instilled into the lining of one of his lungs. This instilled air broke the vacuum that normally allows the lung to expand, thereby collapsing that lung. When my father was in his eighties, he still had vivid memories of the sudden feeling of impending doom that the procedure caused him to feel when he was a young man. He was fortunate that the procedure, first on that one side, and later on the other, seemed to help. For others, the procedure resulted in portions of their lungs being surgically resected, as they became actively infected, with the infections spreading to the less-infected areas of the lungs.

Over time, my father slowly improved. He lost his scholarship when finances became tight at the sanatorium, and he was transferred to a state facility. Amid all this uncertainty and medical wandering, he had the good fortune to meet a young nurse whom—many years later, and when it was clear he would live—he married.

By the early 1950s, new drugs had been developed that were very effective for tuberculosis. My father was given a course of them, the thinking being that, although his disease was inactive, tuberculosis always had a predilection to recur, especially if the patient became globally unwell from other causes. The new antibiotics were so powerful, and the bacteria were so naïve in regard to these drugs (they had not been exposed to them before), that the treatment courses were successful. And my father, who lived to his mid-eighties, did not succumb to tuberculosis. He succumbed to heart failure—no doubt abetted by his smoking cigarettes while bored and wondering if he was going to die in the Adirondack Mountains of New York State.

The bacterium that causes tuberculosis is a clever symbiont of *Homo sapiens*. Genetic profiling has shown that it is around seventy thousand years old, and it seems to have emerged out of Africa, along with humanity. It appears that, in contrast to many other infectious agents, it didn't jump from domesticated animals to man, as seventy thousand years ago there were no domesticated animals. And as *H. sapiens* has become more urbanized, the mycobacterium that causes tuberculosis has also diversified to better survive.

The tuberculosis bacterium has a winning evolutionary package that has enabled it to survive, and even thrive, during this age. It has infected probably one-third of the world's total population, doing best among people living in crowded conditions and people who are malnourished. It has gotten a big boost from the evolution of the HIV virus, whose ability to impair immunity has allowed tuberculosis to flourish.

Many famous people have died from the disease, including authors such as Robert Louis Stevenson, the Brontë sisters, John Keats, Franz Kafka, and George Orwell. Disseminated tuberculosis was called consumption because of the wasting it caused prior to killing you: It "consumes" you. It was also called the White Plague because many of its victims suffered from severe anemia, becoming extremely pale. And, of course, the illness has played a role in the public imagination. Hans Castorp in Thomas Mann's *Magic Mountain* was in a Swiss sanatorium. Mimi in *La Bohème* dies of consumption in Puccini's opera, as does Violetta in Verdi's *La Traviata*. Readers and theatergoers of the time knew well the role—both literal and metaphoric—of the disease in these story lines.

So how does the mycobacterium do its work? One, with rare exception, would acquire tuberculosis by inhaling the mycobacterium from infected droplets—the inoculum—that has been coughed up by a person with active pulmonary tuberculosis. The inhaled droplets are not highly contagious, and, generally, one needs a fair amount of exposure to become actively infected. As you can imagine, an ideal situation for infection would be living in a crowded tenement room with poor ventilation and with one or more actively infected people who are coughing in the room. Once the mycobacterium is in the lungs, the infected person's body mounts a response to it that usually walls it off. This initial response might cause a fever and malaise, or, if the inoculum is not too virulent, there might be no systemic symptoms at all. Once the mycobacterium is walled off in the lung tissue, it stays there until the victim develops a situation in which he can become compromised. This could be caused by the stress of a different infection, or malnutrition, or the ravages of war, or some other circumstance. If the initial inoculum is exceptionally potent, or if the infected person is exceptionally frail, the original infection might spread from the lungs to many other sites in the body. There it might also be contained. Or, if the infection is severe enough, it might never be fully contained and will eventually consume the victim, either as continual lung disease or by interfering with other organs and their functioning.

But most adults who have died of tuberculosis acquired the infection as children and kept it at bay until circumstances allowed it to recur, either in the lungs or in other areas of the body. When tuberculosis recurs in the lungs, it causes cough, decreased appetite, fevers—usually in the late afternoon—and, as it progresses, it can cause erosion of either small or large pulmonary blood vessels. If it erodes into small capillaries, the patient coughs up blood in the sputum. If it erodes into a large pulmonary vessel, the patient rapidly exsanguinates—bleeds out. If the patient had not handled the original infection well, the bacteria could be anywhere in the body before it is walled off: in the lymphatics, kidneys, adrenal glands, bone marrow, meninges, peritoneum, bladder, eyes, or even bones (especially the bones of the thoracic spine). Then, when the patient becomes susceptible and the bacteria are no longer contained, the disease can manifest itself in many ways in many parts of the body. That is why it is one of medicine's great imitators.

Physicians who work in third world countries or other resource-poor settings have seen a tremendous amount of tuberculosis. Many of my colleagues at Brigham and Women's Hospital have done considerable work in Haiti, where tuberculosis is endemic, and they have seen it in all its manifestations. In my practicing lifetime, I have seen it be unexpectedly diagnosed in a patient with persistent and unexplained swelling of the wrist. The biopsy done in the OR showed numerous tubercle bacilli, and the OR had to be rigorously decontaminated. Neither the orthopedic surgeon nor I had any inkling the diagnosis would be tuberculosis. Another case I saw was a woman from India with swelling of her abdomen. The usual causes, like liver disease or malignancy, were excluded. The fluid from her abdomen never cultured the mycobacterium. When her bowel became obstructed, and eventually had to be approached surgically, biopsies of numerous thickenings of the peritoneum lining revealed tuberculosis. At least with this patient, we were all suspicious but could not prove it until our clinical intuition was confirmed by surgery.

The very first patient I ever admitted as a brand-new intern was admitted for failure to thrive. She was an elderly African American woman who had the double misfortune of having a terrible disease and me as her brand-new intern. I am sure I asked her an interminable number of questions. I do remember she kept drifting off as I diligently tried to do all the things I was taught to do with a newly admitted patient. I saw her on the main female ward, interviewed and examined her as best I was able, and then drew some blood from her. As I was in the nearby lab, looking at her white blood cell count, the nurse called me and said the patient was unresponsive. I left the lab, and I remember I spilled the blue bottle of Wright's stain as I hurriedly exited to administer CPR. Until the bag valve mask arrived, I gave the patient mouth-to-mouth resuscitation while the nurse administered chest compressions. In short order, more senior staff arrived to help with the chest compressions and to intubate the patient. Unfortunately, the patient expired in spite of our best efforts. This was a long time ago (1973), and in those days all of us as medical trainees tried to obtain autopsies on our patients. We had been inculcated by our teachers of pathology that it was at the autopsy table that many diagnoses were revealed. My senior resident, a former *National Geographic* photographer who had gone to medical school in his late twenties, was able to speak with the patient's family and persuade them to permit an autopsy examination. As the patient had died soon after I had examined her, and as I did not want the lasting reputation of having caused her death by my feeble ministrations, I sought and received permission to see the autopsy. It was conducted by a truly gifted pathologist, Ramzi Cotran, for whom I developed immense respect. The memory of that autopsy is still clear to me today, as I write these words—his dissection of the organs and his running explanation to me, standing opposite him, as to the cause of the patient's death. Everywhere throughout her body were these tiny, millet seed–sized excrescences, in her liver, in her kidneys, in her adrenals, in her ureters, in her bronchi—and Dr. Cotran told me that, when we examined just one of these nodules, they would be teeming with tubercle bacilli. In short, my first patient as a newly certified physician died of disseminated tuberculosis. She had no fevers and no cough; rather, she just had failure to thrive and a lack of appetite. Her family said she had lost some weight, but that was it. A chest X-ray in the ER had shown signs of old healed tuberculosis, but many people her age had these findings. It was only at the postmortem that the true and disseminated nature of her tuberculosis was revealed. Because of its attack on her adrenal glands, she lost her appetite, her blood pressure was very low, and she was weak and cold. She succumbed to the illness that consumed her by infiltrating so many of her organs. Of course, neither I nor the doctor who admitted her from the emergency ward knew the true cause of her illness.

Disseminated tuberculosis in the United States is now uncommon. It is usually seen in the very elderly or those infected with HIV. It is most commonly seen in people raised elsewhere. One of the reasons it is not seen here commonly is that we now have good drugs to treat the common strains of tuberculosis. Because the tubercle bacillus has a tendency to wall itself off in the body, antibiotics need to be given for prolonged periods of time to have enough time to penetrate the walled-off areas (called granulomas) and to interfere with the slow-moving cell cycle of the tubercle bacillus. The antibiotics initially became available in the late 1940s and were put into widespread use in the 1950s. There was early hope that, with the use of these drugs, tuberculosis might be eradicated.

But that was a very naïve assumption. A bacterium that has survived in parallel with mankind for seventy thousand years would not go quietly into that good night. It soon became apparent that the bacterium, if not eradicated by prolonged and effective therapy, was capable of morphing into a resistant strain or strains. Indeed, in some Russian prisons, where the population has tuberculosis with inadequate treatment, there are strains of multidrug-resistant tuberculosis (MDTB) that are resistant to all currently known antibiotics. When these strains infect incarcerated populations, where spread is relatively easy and access to excellent care is lacking, the disease is relentlessly fatal. Another newer cause of persistence of the infection is HIV, discovered in the early 1980s. HIV, which paralyzes our immune system, makes the host extremely susceptible to either primary infection with tuberculosis or, more commonly, ignites a latent prior infection. Of course, poverty, malnutrition, and crowded living quarters are the long-standing and enabling promoters of the disease.

Like that other great imitator, syphilis, tuberculosis is an infection that continually stalks all of humankind and, on a more individual level, challenges physicians throughout the world.

MOONLIGHTING

I HAD HAD A LONG NIGHT. I WAS MOONLIGHTING in a small fifty-bed hospital in the western part of the state. It was in a blue-collar area with a lot of factory workers. I had worked a Friday-night-to-Saturday-morning shift, and it was the day of the month the factory workers received their paychecks. Often, a significant percentage of the workers' pay ended up at the local bar or liquor store, and I knew it would be a busy night.

I was moonlighting because I had two little kids and a wife who was then a stay-at-home mom. My fellowship salary alone was inadequate to feed my family and pay for the apartment we rented, so moonlighting was necessary to make ends meet—and it was not unusual among my peers for many of them to do just what I was doing. Indeed, the doctor I signed out to on Saturday morning was a young faculty member at another Boston teaching hospital, one whom I had gotten to know a bit because we moonlit in the same hospital.

These moonlighting gigs were passed along like precious gems from friends to friends. I was pleased to work in this little hospital because it was private pay—that is, I billed for the work I did in the little hospital's intensive care unit or in the relatively busy emergency room. I was so young that generating a proper bill, using a billing service, and having that service collect for my bill was exciting to me. I actually got to see the money flow in a few months after I had been up all night tending to the sick and intoxicated. This little hospital mostly served people who had health insurance, so, eventually, I was paid for most of the work I did. Furthermore, the few physicians on the local medical and surgical staff were pleased to have the young folks from Boston come out and admit patients, run the ICU, and take care of the small ER. And, in the not very common event when I needed one of the local surgeons to come in and urgently perform an operation, the surgeon would actually get out of bed and come in. I had worked at another facility for a bit over a year where the surgeons would simply never come in at night, and I thought this was bad care. Of course, I was also working hard to stabilize the patients at those facilities, so that, if they lived, they could be taken to the OR in the best possible condition. This little facility was a good place to work on many levels, and I was happy to take as many shifts as the administrator would give me in a month. The money was good enough that my wife and I started to have enough extra to put into a house fund to save for a down payment on a future home for our growing family.

Back in those days, the discipline of emergency medicine did not exist. Physicians in the ER were mostly young, freshly board-certified internal medicine doctors. We all learned the necessary basic surgical techniques we needed on the fly. Massive trauma was out of our league, and fortunately, the emergency responders knew that the small hospitals were not trauma certified, so they diverted the bad cases to the larger medical centers. But suturing, setting simple fractures, draining ugly abscesses, placing large intravenous lines in people who were bleeding, and intubating people who had overdosed or had cardiac arrests—we did it all. And we did it well. The era of superspecialization—where an orthopedist only worked with ankles and not legs, or a cardiologist only saw patients with heart failure—had not yet arrived. And it was fun. As a trainee in internal medicine, one did not get to routinely do a lot of this stuff. Sewing someone up after removing the shards of a wine bottle from his scalp was fun. Draining a large abscess that had been growing and festering for three weeks until the patient couldn't stand it anymore was also fun—and a bit novel. It made me feel like a real doctor, one who wasn't just trying to persuade people post heart attack to stop smoking and eat less meat—this was a more immediate and gratifying experience. And I learned how to do some things that were more significant and complex.

Early in my moonlighting career, I had a man come in with a collapsed lung. I quickly observed just how uncomfortable that condition is—the patient was struggling to breathe and was very uneasy. Fortunately, a surgeon was in the hospital at the time, and he showed me how to put in a chest tube. It involves carefully avoiding the artery that courses under the rib, anesthetizing the muscles between the ribs and the lining of the lung, making a small hole with a scalpel, spreading the hole with a type of clamp, and pushing a tube attached to a water seal into the pleural space (the space between the lung itself and the chest wall). The surgeon was kind and patient. He emphasized how, with a little care, virtually nothing could go wrong, and he told me, "You'll do the next one by yourself, and if you get into trouble you can call me." About six weeks later, I did need to put in a chest tube for a trauma victim, and it went well, without even the need to call in the kind elderly surgeon.

Of course, with this mostly unsupervised system, things did go wrong occasionally. Although the doctors were screened by the administrators of the moonlighting programs at each hospital, physicians still varied in their ability to go it alone, particularly when things were busy. Doing well in these environments is less about having an encyclopedic depth of knowledge than it is about appreciating your co-workers—the nurses, the radiology techs who often would help you read the CT scan, and the community doctors you would have to call to alert them to an admission to be seen in the morning. The system could break down if a moonlighter simply had a low emotional quotient (EQ)—that is, if the moonlighter did not handle interpersonal relationships well. Arrogance about one's capabilities was another big danger—you could get in over your head in an urgent situation, and if you didn't have the humility and self-awareness to ask for help, you could put the patient in serious jeopardy. Last, dealing with all these different physicians during each shift put hospitals at a disadvantage. The quality was variable among the doctors, and they were not board certified in the procedures they were doing. Also, because the doctors were essentially acting as medical fire fighters (focused on putting out fires and not on looking into the "whys" of the fires), it was a system in which little progress would be made.

An example of this is not catching instances of domestic violence. I saw many, many cases of domestic violence when I was moonlighting. I would rule out a fracture here, or a ruptured spleen there, but I wouldn't see the same couple when they came in again three weeks later with the same issue. I was appalled when I realized I was treating someone who had been battered, but the system then was not designed to see this as an ongoing issue—rather, it was designed to deal with acute problems, moving from one patient to the next patient waiting to be seen. So there is a lot to be said for the evolution of the discipline of emergency medicine. Now doctors are trained to have the skill set they will use. They can deal overtly with the overlap of surgical and medical situations that they see in the emergency room, and a body of knowledge has developed, with evidence-based care now being dispensed to the benefit of the patient. Furthermore, hospitals can rely on board certification in Emergency Medicine to offer themselves and their patients reasonable assurance as to the quality of the physicians in their employ. Because there has long been a shortage of primary care physicians, emergency medicine doctors provide a lot of primary care to patients, and they are now knowledgeable about the social systems that support hospital ERs—systems that place battered spouses directly into safe shelter, or social workers who can place homeless people into safer housing than the housing they were in. And, with the discipline of emergency medicine subject to board certification, research now occurs illuminating the best way to handle medical, surgical, and social emergencies.

But when I was moonlighting, the less-supervised system was a good thing for me. I learned how to handle the ebb and flow of emergencies. I learned a lot about my limitations when I was truly on my own. I learned to dispose of the arrogance that major medical centers unwittingly cultivated in their trainees. I learned that there are many caring, incredibly hardworking medical personnel in many communities, and the fact that they work in small communities should not imply lesser competence or compassion.

And last but not least, I learned how to suck it up: I would work a twelve- or twenty-four-hour shift, go home, and then go to work again. I had to perform at a high level for twenty-four to thirty-six hours straight. Data from the last fifteen years show that I was likely to get into a car accident driving to and fro for having worked those hours, but I never did. And I don't know that I ever harmed somebody while being generally overtired—although that is a fact I cannot prove.

The average medical student now graduates with over $160,000 of debt. This guarantees that an awful lot of them will have to do work beyond their day job to, first, pay off their debt and, second, save money for the middle-class accoutrements—like housing and children—that so many desire. I was fortunate because I left medical school with no debt. Though I went to a private medical school, my parents helped as best they could with tuition, I received some scholarship money, and I was able to moonlight a lot while in medical school. The most enjoyable job I had was going on quiz shows taped in New York City. I found that the producers were happy to put a young medical student on the air, and, having been an English major in college, I actually knew a few facts about history and Shakespeare, which served me well. But the stations were smart: Because they would tend to see the same people over and over again, after being on four quiz shows, they wouldn't let you do more. So I then turned to the less interesting, but steadier, work of drawing blood in the hospital, starting at 4:00 A.M. and finishing by 6:30 A.M. so I would be ready to make rounds.

It concerns me that today medical school debt is so high; clearly, it will influence a young physician's choices as to career paths, as the more lucrative specialties are an easier way to get out of debt. But to me, the good news is that there will have to be many young doctors doing what I did—working all sorts of extra jobs to make ends meet. And I am sure they will learn as much about themselves and about new things as I did when I had to moonlight. And that is a good thing.

PLAYING THE PONIES

I WENT INTO PRIVATE PRACTICE FOR A NUMBER of years. The reasons were multiple, but not the least of which was the difficulty I had supporting my growing family on an academician's meager salary. I was moonlighting one or two nights a week on top of the ten-hour workdays of my regular academic job as a clinician/teacher—a prescription for fatigue and burnout. Faculty salaries in the late 1970s to the mid-1980s were notoriously poor, and it certainly would have helped if I had had a private income or were well-to-do.

My epiphany came when I was signing out the patients I had been covering for the previous twenty-four hours in a small hospital in western Massachusetts to the incoming moonlighter. I had worked from Friday late afternoon to 5 P.M. on Saturday after a full workweek at my academic job. I was eager to be on my way, but my colleague indicated that he had something he wanted to share with me before my departure. He was older than me by about fifteen years, and he was a well-respected Boston academic infectious disease consultant. He told me that this would be the last time we would sign out to each other, as he had taken a job as the chief of infectious diseases at a medical school about a thousand miles away. I liked this fellow; he was thorough, smart, and had a pleasing avuncular manner.

As he told me this, a movie started playing in my own tired head, showing me, in another ten to fifteen years, telling a younger colleague that I would no longer be moonlighting here in this small town, but would be a thousand miles away, finally able to earn enough to buy a home and support my wife and children. Tired, probably a bit depressed, and without a mentor to give me truly good advice—assuming I would have listened to good advice—I decided that enough was enough, and that it was time for me to find a different job.

That part was easy. I found employment in a fine community with an excellent community hospital. But cutting the umbilical cord from academe is not without its costs. When you leave the big city hospital, you not only leave long-standing established collegial relationships, you leave the cutting-edge technology that is taken for granted. Both the intellectual people power and the technologic availability are assets that are intrinsic to doing high-quality work.

One of my responsibilities in my new job as a physician in the community with privileges at the community hospital was to be scheduled as staff physician on call to attend to patients who were admitted to the hospital without having a physician of record. This staff physician model is part of a very old system that devolved from a concept of staff privileges—that is, when you were a physician on the hospital staff, your right to be on the staff carried responsibilities that came with it.

Until the advent of hospitalists—dedicated in-patient physicians who work exclusively in hospitals—community physicians would donate their time to the local hospital to service the patients who were admitted to the hospital who had no physician of record. Usually, these were the poor, disadvantaged, and uninsured people. Occasionally, they might be people who became ill while traveling through the community, but most were locals.

The way it worked is that, when you were the designated staff physician, the local ER would contact you at your office and alert you to an admission you needed to attend. If you were lucky, you could "turf" the admission to a specialist colleague, for example, to a cardiologist if the patient seemed to need the small coronary care unit and specialized care. If you were unlucky, and the patient admitted was very unwell, you had to go to the hospital promptly to assess the patient, determine the treatment plan, and get things moving in the right direction.

This scenario was problematic because you had to leave your office, even if you had patients waiting to be seen, so you could attend to an unknown patient with whom you had no relationship, so you could fulfill your obligation to the local hospital. Another problem was that various physicians fulfilled their staff responsibilities with differing levels of care; some were attentive, and some, not so much. The inequities of this lack of uniformity in patient care for the people presenting to the clinic or ER eventually brought about better methods for ensuring uniform and higher quality of care.

But in the late 1970s, when I was still a new physician in the community, the staff physician model was in use, with all its variability. As a new and younger physician in the town, I was an unknown quantity to the people who, clinically, truly run community hospitals—the nurses. With the responsible physician somewhere in the community but off the hospital grounds, the nurses were crucial in accurately assessing the patient's status and in communicating it to the physician, who was theoretically in charge of the patient's care. Skilled nurses are a pleasure to work with and are committed, capable, and experienced. The best of them are also adept politicians, as they deal with multiple physicians and other staff for all of the patients on their floor.

When I had been in my new location about six months and was the designated staff doctor on call for the current twenty-four hour period, I was alerted to four admissions by the ER at around 11 A.M. A brief conversation with the ER attending physician indicated that three of the four admissions were quite straightforward. There was an older man, a smoker, with a lobar pneumonia. There was a middle-aged woman who was a severe alcoholic, who had run out of money, and therefore vodka, and was now presenting with impending alcohol withdrawal. There was a young newlywed with a high fever and urinary tract infection, probably symptoms of an infected kidney, a condition called pyelonephritis. The fourth patient was a bit of a mystery. He was from a neighboring community, had presented to the ER with a fever and malaise, and did not look well to the ER doctor. The only abnormalities in his labs were an elevated white blood count (nonspecific, and often seen with fever and infection), low platelets (circulating blood elements that are essential for clotting and that originate in the bone marrow), and modest anemia. He did not sound very sick, and the suppression of his bone marrow elements was nonspecific and could be seen with infections, cancer, or a host of other conditions.

I told the ER doctor that, if the ER staff thought it was okay, I would stop by the hospital at the end of my day and start with this fellow when I evaluated the four admissions. The ER doc thought that was fine, and I signed off, getting back to seeing a waiting room full of patients.

About two hours later, I received a call from the head nurse of one of the medical floors. She was calling to inform me that patient number 4, the unknown, didn't look at all well, and I should get over to the hospital ASAP. She hardly knew me, as I was new to the community, and she was in take-no-prisoners mode—she was making it very clear she wanted me there now. When I asked her why she felt this way, she was abrupt and said she thought he was deteriorating neurologically and unless I was also a neurologist as well as an internist, I had better get some help. Apologizing to my patients, who were understanding, I hopped in my diesel Volkswagen Golf and went as fast as that little car could motor to the hospital.

I introduced myself to the patient, S.M., and assessed him. He looked unwell, had a fever up to 102.5°F, and was complaining of a headache. The head nurse, who had come into the room with me, told me she thought the patient had suffered a transient ischemic attack—a stroke—around forty-five minutes earlier, with slurred speech, confusion, and perhaps some left-sided weakness. He had normal speech now and moved all of his four extremities equally. But he was clearly not very with it; he was disoriented as to where he was, who the president was, and how many dimes were in a dollar.

The remainder of his examination was nonspecific except for some purple spots under his skin and some tiny red blots on the skin—both skin findings seen with people with low platelets.

When someone has a fever and is confused, a looming and urgent concern is meningitis. The best way to make that diagnosis is with a lumbar puncture, something that sounds pretty awful to the layperson, but is actually easy to do, and I had done many. However, there are a few situations in which you do not want to ride hell-bent into the subarachnoid space around the spinal column; one such situation is a bleeding tendency. If the patient is predisposed to bleed, the insertion of a needle, even small in caliber, can cause bleeding around the spinal cord, with consequent compression of the cord—a situation where, in an effort to make a diagnosis, you've created an even larger problem. Suspecting that he did have a bleeding tendency, I asked for urgent repeat labs to confirm or deny it and ordered a lumbar puncture kit up to the floor. I could see the head nurse's skepticism about my capabilities increase as I thought out loud.

While we were both standing at the patient's bedside and I was conversing more with the patient in an effort to understand the time course of his illness, he started to vomit. The vomit had a specific coffee-ground appearance, which is a hallmark of digested blood in the upper gastrointestinal tract. This upped the ante even further, as I was now worried that, in addition to possible meningitis, he was bleeding in the GI tract and would need urgent treatment and monitoring for both. I asked if the patient had urinated while on the floor, as there was no report of a urinalysis from the ER, and he had. His urine was dark maroon in appearance, visible in the yellow-tinted plastic bedside urinal.

That visual clue is what made the synapses in my brain suddenly put all his symptoms together. He had a pentad of signs and symptoms that had been described in the literature, though I had never actually seen a patient with the illness. He had fever, neurologic signs (confusion and a transient ischemic attack), low platelets, gastrointestinal symptoms, and anemia.

S.M. had thrombotic thrombocytopenic purpura—TTP. This was in the late 1970s, and the illness was not widely recognized at the time. I had only read about it in a journal that published series of newly described disorders, and, although the illness had been discussed in my training, it was rare enough that I had never encountered it. It was uniformly fatal, and patients tended to die of progressive renal or neurologic disease.

I stepped outside and told the nurse what I thought was going on. She was quite skeptical—here was this junior doctor, telling her first he was worried about meningitis, now changing his tune and citing an illness that he vaguely recalled and knew next to nothing about. Appropriately concerned for the welfare of her patient, she gave me a lot of pushback. To give myself some time to think, I added broad-spectrum antibiotics to S.M.'s orders and figured this would cover meningitis, as well as cover me in the disapproving eyes of the nursing staff.

This was back when there was no Google or even an Internet, and all hospitals had libraries full of journals. Searching took a while; you had to use a publication called *Index Medicus*, find if there were any articles related to your inquiry, and then see if the library you were in had those journals on the shelf. I told the staff where I would be, called the neurologist on staff call, and went to work in the library. I called my office and canceled the patients for the day with my apologies.

I searched in the library for about an hour with little progress. I rechecked on the patient, who seemed no worse, and went to the laboratory to see if the repeat labs were back. They were: The anemia was worse, which may have reflected S.M.'s GI bleeding, and the platelets were lower. I asked the lab technician to make a slide for me of the blood smear, and the slide showed the cells I thought we would find—torn-up red cells called schistocytes.

Seeing the schistocytes confirmed the underlying pathology of S.M.'s condition—he was forming little strands of coagulated fibrin in the very small arterioles and venules of his circulation. These fibrin strands were shearing his red cells, causing anemia, and they were trapping his platelets and consuming them. When these fibrin strands created this tiny spider web of clots all throughout S.M.'s body, his organs were at risk for small clots. As described, the clots mostly tended to affect the brain, kidney, GI tract, and, occasionally, the heart. When the kidney is affected, the urine turns bloody—and that was the visual trigger that made me consider the diagnosis.

The diagnosis was now made, but what to do? I knew that steroids, in high doses, were used in hematologic conditions when there was a sense of an antibody in the blood that caused a cascading trigger of clotting events. I called the floor and started him on high-dose steroids.

But in my search, I saw an experimental treatment of total body plasma exchange. Why or how that worked wasn't clear, and it was expensive and didn't always work. But the abstracts that reported its use were promising. By now the consultant neurologist had arrived, another young attending, to whom I presented the situation. He went to examine S.M. while I continued to pore over the *Index Medicus*. He returned to tell me that S.M. had indeed sustained a small stroke—the findings were more subtle than I was able to detect, and although the neurologist had never seen a case of TTP, he knew a professor in New York City whom he could call on for advice. He did so, and the professor, who freely gave us his time, said he had experience with plasma exchange and that it had helped a number of his patients at his major New York City medical center.

I got on the telephone and started calling to see how plasma exchange (called plasmapheresis) could be done in our community hospital. We found out that the machinery had to come from seventy miles away with a technician and would arrive, at best, in around eight hours.

Both the neurologist and I continued to observe the patient. We gave him blood, followed his declining kidney function, stopped the useless antibiotics, and watched helplessly while he sustained a significant stroke that caused his speech to become garbled. Those trying to help S.M. suffered what is too common in medicine and rarely spoken of—the situation in which, sometimes, all we can do is stand by the patient with hope, expertise, and compassion while observing a disease do its worst.

By late evening, the machinery had arrived. The technicians had to set everything up. A surgeon had to be called in after hours to place a large-bore access line. Late into the night, I watched the plasma exchange machine exchange S.M.'s plasma for plasma from donors. After six hours, repeat lab testing showed a marked increase in his platelets and stabilization of his anemia. With the advice of consultants, both in the local community and in major academic centers, S.M.'s plasma exchange went on for a week. Within twenty-four hours of his being admitted to the hospital, his situation was stable enough that I felt I could leave his bedside and start seeing the next day's patients as well as the patients I had canceled earlier. That evening I went home and was asleep for the first time in forty hours.

The good news is that S.M. survived. His kidney function, which had been acutely destroyed, slowly recovered. He required dialysis for over six months. His neurologic deficit, however, never improved, and he was left with a weakened arm and leg and garbled speech.

The other good news is that my neurologic colleague and I became close and fast friends from that long evening and night at S.M.'s bedside, as we watched, helplessness, as the disease progressed unchecked. The head nurse developed a grudging respect for my abilities, and one day, after her own doctor had retired, I was pleased and honored to welcome her to my own practice.

And there is a final human piece to the story.

S.M. had decent health insurance, and I sent him a bill for the considerable time and effort spent on his care. My office manager told me that, around eight months after we had submitted the bill to his insurer, the insurer had sent the payment to S.M., but he had never paid us. S.M. had followed up in the office with me and with the neurologist. On his next visit, I brought up the uncomfortable situation of his outstanding bill. His wife told me they had cashed the insurance check, and both had gone to the track and played the ponies. They figured that S.M.'s life was now pretty lousy and that I had enough money. And they never returned.

WHO'S THE GREATEST OF THEM ALL?

FRIENDS AND FAMILY OFTEN ASK ME WHO THE best internal medicine doctors are. I never hesitate to respond: infectious disease doctors. My opinion is, of course, fashioned from the people I have encountered and the places I have trained and practiced, but I would suspect that across the country, and perhaps even throughout the first world, this endorsement would endure.

I base my opinion on a number of factors.

Infectious diseases take all shapes and sizes: They encompass the well-recognized rash and pain of shingles (herpes zoster); the urgent and life-threatening case of bacterial meningitis; the black little eschar on top of a skin ulcer of cutaneous anthrax; the enlarged, exquisitely tender lymph nodes of bubonic plague; and the rose-colored spots of typhoid fever. These and many more syndromes are recognized in their entirety by these specialists, who are current as to the most appropriate therapies for all infections that are amenable to treatment. In the hospital where I work, one infectious disease colleague's job is to monitor the microbiology laboratory and see what strains of bacteria are becoming resistant to the antibiotic standards of care. He generates these data every day! Another colleague who has spent time in Brazil studying Chagas disease is an expert in parasitic infections from around the world. A third colleague treats and studies HIV patients in our hospital, and his treatment plans are so refined, and so compassionate, that he is a role model for many of our young trainees.

So far, I've been talking about the intellectual prowess of these individuals. But my judgment in rating them as truly the best doctors is more nuanced. All of them—and I believe this is true of their peers in many other sites—are truly fine internists and compassionate physicians. They are nonjudgmental, they are always ready to give of themselves, even when the patient is very difficult or refuses to follow a treatment plan. Though many of them have experience with illness in exotic locales or the third world, what drove them to these places was a desire to help the underserved and the disenfranchised.

As I write these words, there is an entirely new generation of physicians who are not familiar with the unchecked ravages of HIV/AIDS. The illness, which was first widely recognized in the early 1980s, was truly devastating in its effects on both individuals and caregivers. The infectious disease consultants asked to manage these patients had very little to offer beyond their compassion and willingness to stand by them. As John Milton said, "They also serve who only stand and wait." Early in the epidemic, I tried to manage the two patients I had with HIV/AIDS. All of us were still learning about the illness, and I was reluctant to surrender the care of these patients to specialists who didn't have longstanding relationships with them—but I was wrong to have this attitude. Both of the patients ended up with intractable diarrhea from a parasite and simultaneously developed a severe wasting disease. One day, one of the two patients had severe seizures due to a brain infection, in addition to all his other problems. My infectious disease colleagues helped me treat both patients with the agents available then—agents I had no experience in prescribing. As their illness continued with little abatement, I spent more and more time with both patients and felt more and more helpless as they continued to suffer from both their primary illness and the opportunistic infections that exhausted them and stole their lives. Seeing how much my very small sample of patients suffered, I developed tremendous respect for the infectious disease consultants who had many of these patients. I never heard any of these physicians make judgments about the patients' lifestyles or personal choices, their ability to pay for their care, or the demands their illnesses made on physicians' time.

Furthermore, patients with HIV/AIDS often had occult medical problems that required a very sophisticated and high level of diagnostic acumen. Of my two patients, one was a middle-aged woman who had received multiple blood transfusions during a complicated delivery in another city. The screening for transfusions in the late 1970s was not anywhere near as sophisticated as it is today. She became infected with both the HIV virus and the hepatitis C virus. She developed HIV/AIDS and a wasting syndrome from it. She also suffered from a parasitic infection in her GI tract that caused severe diarrhea and malabsorption. It became clear, with no successful therapies, that she was going to succumb to her illnesses. Too weak to get up from her bed, she often soiled herself in the hospital bed, to her shame and embarrassment. Close to her end of life, my infectious disease consultant called me and said, "I think she has liver cancer." I was puzzled, as she had enough medical problems without invoking yet another. He replied that when he examined her, she appeared jaundiced, and he thought he felt a new mass in her liver. He then listened over her liver and thought he heard the sound of blood flow in the mass—liver tumors are often vascular. He was right, as was proven at autopsy when she had expired six days later.

My point here is that my colleague, even knowing that this patient was terminally ill, was concerned enough, and thoughtful enough, to not hurry away from her bedside but to examine her, think about her condition, and suggest new possibilities. I myself was long past entertaining anything new for this poor and unfortunate woman and was trying only to muster the courage to see her every day, keep her as comfortable as the staff and I were able, and offer honest counsel.

These days, fortunately, we have ways to address an HIV/hepatitis C combination. An infectious disease specialist would see the patient, know that HIV and hepatitis C confer additional risks to each condition, and would tailor a treatment plan to manage both viral conditions. As long as the patient is compliant, the hepatitis C is likely to be cured, and the HIV is likely to be chronically suppressed. Furthermore, the consultant would tailor the various HIV medications to best optimize hepatitis C treatment, to minimize the side effects of the drugs (such as elevation of cholesterol and triglycerides, or bone loss, or the difficulties of having to take multiple pills two or three times a day), and to help ensure compliance with the treatment program to lessen viral resistance.

The evolution of treatment for both HIV and hepatitis C in the last forty years is a remarkable testament to bench-to-bedside research. From the basic understanding of the mechanisms by which these two viruses invade cells and avoid elimination, researchers were able to develop drugs that interfered with how the viruses attach to, replicate within, or hide in host cells. The achievement of sustained cure in hepatitis C is one of the major milestones of medicine in the twenty-first century. Although HIV currently can only be suppressed, not cured, there is hope that a cure will be found or a vaccine will be produced that will prevent transmission. Large international pharmaceutical companies have produced these wondrous results, and there continues to be a national debate as to how to justly monetize these achievements of the private sector. But none of the results of this basic research would have been successful if infectious disease specialists hadn't helped in the basic research, in the clinical trials, and by their advocacy for patients suffering from these illnesses.

A similar story could also be told of the evolution of medicines to treat cancer and rheumatic diseases, and the work of oncologists and rheumatologists as superb clinical, caring physicians who now have this armamentarium of drugs available to them should not be overlooked. But in my lifetime, in seeing the suffering caused by HIV/AIDS and the compassionate care and diagnostic acumen of the specialists who have helped contain it, I rate infectious diseases as the specialty with the most remarkable physicians.

MAKING A LIST AND CHECKING IT TWICE

MR. R. WAS OBESE. I MEAN, REALLY OBESE. Obese to the point he had to use the only plus-sized MRI machine available in Boston. His wife, Mrs. R., was a doctor lover. She never was satisfied with just a second opinion. For every hangnail, freckle, or single change in bowel movements, she wanted three or four opinions. She was the captain of the marital ship and would be in the room whenever I talked to Mr. R. She remained in the room when I examined him (even in the most private of regions), and she would talk for Mr. R. when I was summarizing my thoughts about Mr. R.'s visit.

There is a maxim in medicine that I don't fully endorse but that is sometimes useful to remember: 50 percent of the time, what brings the patient into your primary care office will go away with no intervention on your part. This includes obvious things like a cold. But sometimes even puzzling and unusual things that pop up in a patient's lab findings or symptoms will, if left alone for a bit, just go away.

Mr. R. had some odd things pop up. For example, he once had mild anemia. It was not very abnormal, say a point below the usual cutoff for his age, so I suggested we just follow up in a month, and, lo and behold, in a month he would be back to being within the normal limits.

One day, Mr. R. and his wife came to the office, and Mr. R. volunteered that he was feeling unusually fatigued. Mrs. R. was quite concerned because they were due to celebrate their fiftieth wedding anniversary in a few months in Las Vegas. All the children and grandchildren were coming, and Mr. R. had to be feeling okay by then. I was on the clock to get to the bottom of this.

Because Mr. R. so often had a complaint, or a lab value, that was evanescent, I was happy to take a detailed history, do an exam, and hope once again that within a month the issue would resolve itself—and that the next office conversation would be about which hotel in Vegas the family had chosen.

When someone complains of fatigue, an internist is never surprised. It is the third most common primary care complaint after low back pain and hypertension. The causes are legion, and experienced physicians try to minimize pursuit of the rare and occult by taking a careful history and then completing a careful examination. Fatigue as a complaint can run the gamut of causes from psychological stress to terminal cancer with all the stops in between.

In Mr. R.'s situation, he truly felt fatigued; that is, he felt exhausted all the time, had little energy, and had difficulty finishing his shift as a part-time pharmacist at a nationwide drugstore chain. He denied depression or change in mood, and the ever-present Mrs. R. confirmed that Mr. R. was content to sit in his extra-large chair at home and watch TV and laugh along with the laugh track as he usually did. He was not an exerciser, and, as he was morbidly obese, he did not do more than walk to his car from his house, walk into the store where he worked, and stand or sit on a stool for most of his eight-hour shift. He had noticed no signs of bleeding from his gastrointestinal tract or urinary tract. He continued to eat the same things and in the same amounts. He had no fevers, swollen glands, pains in the lungs, abdomen, or vascular tree (as all the blood vessels combined are called). He did have severe osteoarthritis of both knees, and this continued as before, no better and no worse.

Mr. R. took a host of medications. Modern electronic records pay a lot of attention to medications, and rightfully so. Before we could keep medications all in one place in a computerized chart, it was too easy to lose track of them: when we needed to remember when they had started or stopped, or when we needed to determine which ones had caused some untoward symptoms. Mr. R. took over thirteen medications daily, with another eight on an as-needed basis. He saw a lot of specialists (at his wife's insistence), and many of his specialists were at locations other than my hospital. So I did my best to ask Mr. R. at each visit whom he had seen, for what reason, and what, if anything, had been prescribed. And then I would update his medication list in the electronic record.

The medication review is important at each visit, but especially so when the patient is complaining of fatigue. Medications, even if taken for years, can cause side effects or drug interactions that might well be missed if not reviewed specifically. I went over all of Mr. R.'s meds, about half of which I had prescribed, and nothing seemed untoward or causative of his symptoms. One of his medications belonged to a class of cardiovascular medication called beta blockers, and these can cause fatigue, but Mr. R. had been on the medication for seven years, and it was helping his blood pressure. I was willing to discontinue it, but only if no other cause was found. Because he had severe osteoarthritis of his knees, Mr. R. was taking a nonsteroidal anti-inflammatory drug daily. I was hoping that this drug was the cause of some silent GI irritation and oozing of blood from the GI tract, his past mild anemia, and, if significant, his fatigue.

But no, his anemia wasn't present. However, his labs showed a slight change in his sodium and potassium. His sodium had dropped to below normal, and his potassium had bumped to the upper limits of normal. These modest perturbations should not have caused fatigue, but they were worthy of investigation.

When sodium drops, potassium goes up, and, all other things being unchanged, one has to think about a deficiency of the hormone cortisol as the cause. Cortisol is manufactured in the tricorn hat–shaped glands that sit atop of our kidneys, the adrenal glands. Like many hormones, cortisol's actions are widespread throughout the body, and it is essential for our body's wholesome functioning. People who are deficient in cortisol experience fatigue, a sense of being unwell, GI pain, weakness, and low blood pressure, and they are often quite slender and have hyperpigmented skin. The easiest picture to mentally call up is that of the young senator John F. Kennedy: very slender, always tan—he had adrenal insufficiency.

Mr. R. was anything but slender, and he wasn't hyperpigmented. But stranger things have happened in my practice than an obese middle-aged man having adrenal insufficiency. And, like many hormone deficiencies, they often creep up on both the patient and the physician. So the test to do was a morning cortisol test. Cortisol levels are at their highest in the early morning, and by late afternoon, they start to get quite low. The adrenal glands really get to work just before we normally arise in the morning, and, by the time we are up, dressed, and having breakfast, they have worked up a good head of steam. So, if you suspect that they aren't working properly, the early morning is the best time to draw blood for the test. I told Mr. and Mrs. R. that I was going to do this test, knowing that Mrs. R. would be on the Internet in a New York minute and that she would be suggesting, in two New York minutes, that Mr. R. see an endocrinologist. As I was groping a bit here, I wasn't eager to send Mr. R. to a specialist without confirmation that this indeed was the cause of his fatigue and new electrolyte disturbance. I could imagine the endocrinologist seeing this obese, overmedicated, and over-consulted man in his office for a bogus diagnosis—not something I wanted to be a party to.

Surprisingly, the morning cortisol test was very low. Huh! Here was this man, looking like the opposite of the poster child for low cortisol. Whenever a physician is confronted with an endocrine abnormality, the testing procedure is routine: If you suspect failure of an endocrine gland, you try to stimulate the gland. If the gland does not react, you then sort out, by testing, whether the problem is with the gland itself (the gland is not functioning properly) or with the pituitary part of the brain, which sends signals to the gland to maintain proper levels of the hormone in the bloodstream. Conversely, when hormone levels are too high, you see if the gland is able to suppresses its hormone levels. If it doesn't suppress, it is acting autonomously, either because the gland itself has gone haywire or because the pituitary or another source of a driving force to the gland (like lung cancer) has escaped the normal feedback mechanisms.

Because Mr. R.'s morning cortisol level was low, the next step was to stimulate his gland with synthetic pituitary cortisol-stimulating hormone. The test involves drawing baseline morning cortisol, giving the patient synthetic pituitary stimulant, and seeing in an hour if the gland responds appropriately. If the response is minimal or muted, it means the gland isn't functioning properly. Sure enough, Mr. R.'s response was very poor.

I double-checked his medications. I reexamined him. I questioned him again. He wasn't taking any medication that could interfere with his adrenals. His fatigue was worse, but his electrolyte abnormalities were about the same. He still experienced GI discomfort.

It's as true of medicine as it is of poker: You have to know when to hold 'em and know when to fold 'em. I couldn't figure it out on my own. I called a colleague who is an esteemed endocrinologist. I asked him to see Mr. R. and told him that I was puzzled. As often happens in our hospital, he carefully listened to me describe Mr. R.'s story, and he promised to get Mr. R. into his office within a week.

And the specialist provided the answer. Mr. R.'s adrenal glands were indeed suppressed, and the cause was his medications. Mr. R. was seeing a couple of orthopedists, neither of whom knew the other was seeing him. Both were giving him timed-release steroid shots to his osteoarthritic knees. When I was seeing Mr. R., the timed-release effects of the shots had mostly worn off, so the effect of the steroid injections was no longer present in the bloodstream. But their long-term effects caused suppression of Mr. R.'s adrenal glands, and that was what I was seeing. The endocrinologist told Mr. R. to receive no more steroid injections for six months and that he should inform his primary care doctor (namely, me) about all of the physicians he was seeing and what they were doing.

In about three months' time, Mr. R.'s adrenal glands were mostly back to normal. His fatigue was much improved, and his electrolyte abnormalities were normalized. His mild anemia (another likely effect of his suppressed adrenals) was gone. He still had bad osteoarthritis and would certainly need knee replacement surgery (but I was hoping he would be able to lose some weight before this surgery). His problem, though, was an old one—medication overdose and lack of communication among all of his doctors, causing unexpected side effects, a major evaluation, major investments of time and money, and major patient consternation.

POEMS

MR. F. WAS A POSTAL WORKER. HE COULD
have done a lot of jobs, as he was intelligent, read widely, and
was an honorably discharged Vietnam veteran. But, whether
it was because of his upbringing with a physically abusive
father or his experience being "scared shitless" in the tun-
nels of Vietnam, he wanted a safe and easy job. So after a
few years driving long-haul trucks, he passed the U.S. Postal
Service exam and became a letter carrier.

He was well liked by the people on his route—and I was
one of them. He was cordial, but not overly friendly. He was
quiet and had a thoughtful air about him on his deliveries.
Though I had encountered him many times on Saturdays
when he made his mail delivery (I am always hoping for some
exciting thing in the mail—a trait my wife finds hilarious), I
was surprised to see him one day in my waiting room.

I hadn't known his last name, and when I saw Mr. F. listed
as a new patient one day, I did not place him as my mailman.
I hadn't seen him in a few months, and when I greeted him in
the waiting room to escort him back to my consulting room, it
was obvious he had lost some weight. When he rose from the
chair in the waiting room, I saw that he had some trouble walking, and I thought he might have weakness in one of his feet.

He was divorced, had two children with whom he
had good relations, and did not have any emotionally or

financially frustrating interactions with his ex-wife. He had grown up in the upper middle class, and his family had left him some money. But he enjoyed his work as a letter carrier, and he used the money in excess of his salary for trout fishing vacations in the American West or in South America when Boston was freezing.

He quit smoking around fifteen years ago and had been a moderate smoker before then. When he served in Vietnam, which he did right after prep school, he had smoked a lot of marijuana and for a while had shot up heroin—less to get high than to cope with the fear he and his platoon mates felt on a daily basis. When he returned home, he couldn't decide if he wanted to go to college, though his family expected he would. But he was always a little contrary, and so he didn't go. It was only after his job as a trucker and his marriage that he went to the University of Massachusetts Boston at night while working days as a letter carrier. Except for his divorce, he had a stable, solid life. He only rarely smoked marijuana and did not drink alcohol because he never liked it. He lived alone and cooked for himself or ate out, and he had good knowledge of healthy eating. He did not do any additional exercise, but he walked around 60 percent of his route, and he figured that gave him sufficient physical activity. All in all, his life was on an even keel, which was the impression he gave when I used to encounter him at the mailbox.

He came to see me because he had delivered mail to my home for years and saw the journals and other medical mailings I received. He told me that he had asked a few neighbors on his route about my competence, and many had said that they had heard I was a capable physician. This kind of referral pattern—asking acquaintances and neighbors, or looking at online reviews—is the way most people find their physician. But it is not a very good way to vet a doctor. The people who write online reviews of doctors are frequently disgruntled patients who can vent easily online. Neighbors rarely know much more about their next-door neighborhood doctor than they would know about their next-door neighbor's snow blowers when they ask to borrow them because theirs has broken. And Mr. F.'s showing up in my office was truly not very likely statistically because probably 20 percent of the people on his route were doctors—I live in a sort of "Pill Hill" section of my town.

He came to see me because of mobility problems. He had developed numbness, tingling, and a sense of coldness in both feet over the last eight months. He had not thought too much of it; it was an inconvenience, occasionally the tingling was actually painful, but it was by no means incapacitating. However, in the last five or six weeks, he had noticed the subtle inability to lift his right foot. It was not paralyzed, but it was quite weak. He had to be careful, as he quite readily would trip over his right toes. Within the last two weeks, the same thing seemed to be happening with his left foot. That really got his attention, and he decided to make a doctor's appointment.

On questioning him, I found that he had no rashes, but he did think the skin on his extremities was a little more hirsute than it had been. He also thought that the skin around his armpits was a little darker than it had ever been. He knew he had lost weight, he guessed around ten pounds in the last three months, because his pants were loose and he had developed a little wattle on his neck. He also felt that his energy had decreased—he wanted to sleep more—and he thought his enthusiasm for his pleasures, namely fishing and the Boston Red Sox, was diminished. Otherwise, extensive screening for other symptoms and signs were negative.

When I examined him, he was certainly correct about the dark hyperpigmentation under both armpits. I could not tell if he had more hair on his legs and arms than he had in the past, but he certainly was hirsute. He had no swelling of his legs. Heart and lungs sounded normal. I wondered if he had a slightly enlarged spleen, but there is no great way to assess this by physical examination, so I could not be certain. His pulse was low—in the low 50s. His temperature was slightly low, and his blood pressure was low for his age. But his neurologic examination was floridly abnormal. In testing the sensation in his feet, I found he could not feel the vibration from a tuning fork applied to his big toe or anklebone. He also could not feel a cotton fiber wisp applied to his feet and moved up to his mid-shin. The muscles that extended his foot were very weak, more so on the right side than the left, and it was clear that he had trouble raising his right foot even a little.

For a man in later middle age to have weight loss, fatigue, neurologic findings, and perhaps organ enlargement causes all kinds of red flags go off in the internist's mind. And none of those flags represents good news for the patient.

I told Mr. F. that I wasn't sure what was going on, but that we would start simply with some blood testing. Those tests would be back in twenty-four hours, and we could talk after office hours the next day. Then, depending on what the blood tests showed, I might want to image his abdomen, as I suspected an enlarged spleen, and I might want to refer him to one of the neurologists in my hospital. I told him that I knew for sure that his peripheral nerves to his feet were damaged—possibly irreversibly, but hopefully not—and that there were many causes of peripheral neuropathy. I told him I would expeditiously try to get to the bottom of things, and if I was stumped, I assured him that I would get him seen promptly by some of the many consultants that worked in my academic teaching hospital.

Mr. F. absorbed all this calmly, was happy to proceed with the lab testing, and was willing to call after office hours the next day on the direct line I gave him.

When physicians are confronted with an ill patient, they try to figure things out efficiently and economically. Another way of putting it is that we don't use a cannon when a BB gun could give us the answer. Part of that drive for efficiency engages the physician's brain as she elicits a history and does a physical examination tailored to the questions the patient's history raises. This dual process of listening and simultaneously formulating hypotheses about the diagnosis is an acquired skill. It draws on the physician's knowledge base, communication skills with the patient, and the physician's ability to properly examine the patient. Other factors interplay with this process, and they have to do with how physicians think.

If a doctor has just seen three documented cases of strep throat in the last twenty-four hours, he is likely to think that the next sore throat he sees in the office is also strep—but statistically it is more likely be caused by the much more common cold's viral diagnosis. Similarly, physicians may have all kinds of biases about what a patient might have, based on their own assumptions about socioeconomic class, ability to articulate a problem, presence or absence of addiction and the associated drive for medications, and a host of other factors. All of these "shortcuts" in our thinking can help provide us with a likely answer, but, in more complex cases, they can actually get in the way of our having a mind open enough to put all the facts together correctly. Last, in making a diagnosis, rarely do all the facts fit the way the textbooks idealize things. One of the most difficult things for trainees in internal medicine to grasp is the degree of uncertainty that underlies all experienced clinicians' diagnostic hypotheses. Television medical dramas never really emphasize how doctors use facts or how they mentally acknowledge what "fits" and what does not "fit" as they come up with working diagnostic hypotheses. And honest doctors are ready, promptly, to revise their thinking as new information becomes available and the priorities of the weighted information change and cause a new, often unrelated, diagnosis to emerge.

What I was thinking about early in Mr. F.'s description was his past heroin use while serving in Vietnam. Many such people acquired hepatitis C infections. Hepatitis C can be slowly and mildly active in some patients for years. It can then cause cirrhosis and occasionally even liver cancer. It can also cause a condition called cryoglobulinemia, in which the body makes a lot of a large serum protein that precipitates out of solution—turns solid and clumps in the blood—when it's cold outside. These cryoglobulins can damage peripheral nerves and cause the signs and symptoms Mr. F. had in his legs and feet. If his spleen were enlarged, it could be from cirrhosis of his liver—so that was a supportive finding.

But some things did not fit, such as the definite hyperpigmentation. And Mr. F. also thought he had increased hair on his extremities—that, too, I could not readily throw into the mix. I knew I needed liver function tests and a test for cryoglobulins. As I suspected long-standing hepatitis C, I ordered the standard antibody testing for that condition. His fatigue, low pulse, and low temperature were all consistent with diminished thyroid function, and the test for that is easy to obtain and is very sensitive. In a general workup, we always get a complete blood count to look at the bone marrow's blood-forming elements. They are often affected when the patient becomes systemically ill, and analysis of all the blood-forming elements often gives clues as to what is going on. There is a nonspecific indicator of inflammation, called hs-CRP, and I ordered a test for this, too, as I thought it would be very abnormal. Last, hepatitis C can cause kidney disease, so I ordered kidney function tests and a urinalysis.

Each day, I kept a checklist on my office desk of patients whose lab tests or X-rays I wanted to look at, and Mr. F. was my number 1 patient for that day on my lengthy list. There was no hepatitis C, and liver tests and kidney function were normal. He had a very high value for his inflammatory marker. His red blood cells, which are part of the blood-forming elements, actually showed an increase (often with inflammatory conditions, the red blood cell count diminishes), and his clotting elements, called platelets, also increased. His supersensitive thyroid test showed that he had a moderately low functioning thyroid, which is easy to address and correct, but it did not explain the entire picture. I told Mr. F. that once we had a handle on what was really going on, it would be very simple to fix this thyroid piece of the puzzle.

I told Mr. F. what these values were, and I had to admit that my leading diagnosis was not tenable. I told him that I wanted to get a CT scan of his abdomen to confirm or deny an enlarged spleen, as other organs can enlarge and "hide" in the abdomen or the region behind the abdominal organs. I also told him that I would speak with a neurologist the next morning to try to get him seen for his neurologic abnormalities.

Mr. F., I could tell, was a bit dismayed. Perhaps I had been a little too sure of myself when I discussed the leading possibilities with him, and now he knew that I had been wrong. I assured him we would get to the bottom of this. I told him I had to rethink the whole situation, and I asked him to have some forbearance, as this was our first encounter. I knew he didn't know anything about my methods or even if I was a capable doctor.

Texting and driving is now against the law in most states, but there should be a law that doctors are not allowed to think about their puzzling patients while driving—because that's what I do all the time, and it is dangerous for other drivers! By the time I arrived home, I had dredged up an idea from somewhere very deep in my cortex. I had read about a condition called POEMS many years prior. The name is yet another medical acronym, one among so many that are frequent in our work, and it stands for "polyneuropathy, organomegaly, endocrine abnormalities, monoclonal protein abnormality, and skin changes." I had never seen a case, but it was rare enough that it would be mentioned in case conferences when no one knew the diagnosis. Polyneuropathy is a description of what was wrong with his lower extremities. Organomegaly means organ enlargement (I thought of his spleen). His thyroid, which is an endocrine organ, was not functioning normally. The monoclonal protein—an abnormal protein sometimes found in blood—was, at that moment, unexplained. The skin abnormalities could well be the hyperpigmentation and excess hair. I wasn't familiar with the POEMS syndrome, I simply remembered it existed. So, to my wife's frustration, when I got home, I took off my jacket and ran upstairs to my study, where I looked it up. My reading told me that, if Mr. F. had certain proteins in his blood called lambda light chains, or if he had some abnormal bony lesions, these findings would cinch the diagnosis. And, even better, there were reports of the neurologic abnormalities being improved when treatment of the condition was started promptly.

POEMS was indeed what Mr. F. had. He received the appropriate treatment by my hematology and radiation therapy colleagues, and he is currently in remission. I no longer have the pleasure of seeing him deliver my mail, as he went out on disability and then transitioned to retirement, but I still see him in the office, and we chat about the neighborhood every time we get together.

IRON MAN

MR. B. LIVED TO SAIL. OF COURSE, HE LOVED his wife and kids, but if you forced him to choose which one thing he could love, it would be sailing.

He had grown up on the south shore of Massachusetts, and from an early age he had been on sailboats. His family owned the local large hardware store and lumberyard, and the family business connected him to his many wealthy neighbors. When he was in his teens, he served as crew on boats sailing in the Bermuda race from Newport, Rhode Island. And when he was in college, earning a teaching degree, he took a year off to crew in a boat that sailed around the world.

I met Mr. B. through a colleague of mine who was his childhood best friend. Mr. B. was referred to me when his prior internist retired. He was healthy, in his early fifties, and in charge of the family business, which was doing fine. His children and wife were all well and were all sailors, so they did not begrudge Mr. B. his time on the water in his off hours. Mr. B. would mostly sail locally, racing in his town's harbor every Wednesday evening in the summer and in some nearby weekend races off Newport.

Mr. B. was a vigorous guy, and I had known him for about six years before he actually came to see me with a complaint. And when he did, he had two complaints. First, his hands hurt him. Second, he felt that he had less energy. There was no family history of arthritis, and Mr. B.'s complaint wasn't that his joints were swollen; rather, they ached. But he used his hands a lot and was always pulling on ropes when he was sailing, so I simply figured his hands had aged about fifteen years beyond his true chronological age. Examining his hands did not reveal anything beyond callused palms and only slightly tender joints.

As to the fatigue, he worked long hours, played hard, and was buying another lumberyard in a neighboring community, so he was sleeping little. I did the usual laboratory tests when someone complains of fatigue. He was not anemic, his liver chemistries were normal, and his thyroid function was fine. The only abnormality on lab screening was a minimally elevated blood sugar (glucose) level, tested at 3:00 P.M., and that was not enough to account for his fatigue.

We agreed to do no more investigations at this time; rather, we would see, after the lumberyard deal closed and sailing season was behind him, if his symptoms didn't improve. If they didn't, we agreed that we would peel more layers of the diagnostic onion. I dispensed the customary internist advice—get more rest, eat less refined sugars, and lose around ten pounds by eating prudently—while hoping that Mr. B.'s complaint would resolve itself on its own.

I had seen Mr. B. in July, and he returned to see me in late November, about three months later. He had felt less fatigued in the interval, but he did not feel his normal vigorous self. His joints still bothered him, but less so now that he was sailing less. And he was taking large amounts of an over-the-counter nonsteroidal anti-inflammatory drug called naproxen. Unusually, he had taken my advice to heart, gone on a prudent diet, and lost around six pounds. He and his wife had gone on a late summer vacation to Florida, and the only thing that had spoiled it was that he became quite ill after eating some oysters and developed severe diarrhea. It persisted for a few days, so he went to a Florida ER, was treated with an antibiotic, and in a few days' time the diarrhea had resolved. The lumberyard deal had gone through, and the stress level from dealing with merged personnel, banks, and the prior owners was now fading.

Once again, Mr. B.'s physical findings were minimal. His knuckle joints were a little tender but not swollen. His heart and lungs were fine, his abdominal exam was unrevealing, and his neurologic examination, including an assessment of his muscle strength, was normal. I noted that he had even seemed to have kept his tan from his Florida trip.

Lab testing once again revealed an elevated blood sugar, and this time it was elevated enough for me to be a little concerned that Mr. B. was developing adult-onset diabetes. My concern was heightened by the fact that he had gone on a healthy diet and had lost weight, but his numbers were worse. Again, I did not think they were bad enough to cause fatigue, but we would have to concentrate on improving the values, or he would have to go on medication to help reduce his blood sugar. This time, though, his liver enzyme tests were slightly elevated. Mr. B.'s alcohol consumption was truly modest; at most he had one beer five times a week (for a total of five beers), and that was the case even when he was involved with evening racing. So, the elevated liver test was a surprise. Sometimes, when blood sugars get quite high, the liver can become stuffed with fat, and this, in turn, can cause liver tests to become abnormal. But, again, I didn't think the degree of his glucose elevation was sufficient to be causing a fatty liver. I wondered if the infected shellfish had caused viral hepatitis on his vacation.

We had a lot of follow-up testing to do. I told Mr. B. what was on my mind, and he reminded me that I was forgetting about his arthritis complaint in his hands, so I scheduled some hand X-rays. There are a lot of joints in each hand, and hand X-rays are inexpensive and easy to obtain, so I sent Mr. B. off to the lab for more blood work and the hand films. I told him we should talk the next day after I had time to review his labs and X-rays.

The hand X-rays revealed the unifying diagnosis.

Mr. B. had a disorder of iron homeostasis called hemochromatosis. Basically, the term means "full of iron." His hand X-rays showed the classic squaring of his third and fourth knuckle joints with a little hook on the bottom that is really only seen in this condition. Furthermore, though the X-rays gave the clue to the diagnosis, hemochromatosis also explained his elevated liver tests and blood sugar, and it was likely the cause of his fatigue. What I did not realize then, in the "aha" moment that occurred when I saw his hand films, was that hemochromatosis also predisposed him to get an infection from the *Vibrio* organism carried in some shellfish.

We normally ingest adequate iron in our American diet. Iron is naturally found in two forms. One is in the blood of animals, and this form of iron is readily absorbed by our bodies. The other is from plants, but there is less iron in plants, and it is less readily absorbed, so vegetarians can sometimes have a negative iron balance. Daily iron losses occur through the skin, when dead skin cells are shed, through the normal wear and tear of the cells of the gastrointestinal tract, and, in women, via menstruation and lactation. If we start to become iron deficient, our bodies know this, and we absorb iron more readily. The inverse is also true: if we have excess iron in our bodies, we absorb it less avidly from our GI tract.

Mr. B.'s illness, hemochromatosis, causes an abnormality in the body's regulation of iron. People with the condition just continue to readily absorb iron, even if they already have more than adequate iron stores. The cause of the condition is a genetic abnormality on the human chromosome 6. If people have two *C282Y* alleles of the *HFE* gene, they absorb iron daily without ever reducing the amount absorbed. Over many years, this iron gets slowly but surely deposited in the cells of our bodies. For reasons not entirely clear, the iron is deposited (and causes damage to) the insulin-producing cells of the pancreas. The iron also gets into liver cells and, if unchecked, can cause cirrhosis. Iron also is deposited in the pituitary gland, which is the master gland that sends messages to the thyroid, adrenal glands, and gonads.

In Mr. B.'s illness, his elevated blood sugars and elevated liver tests were caused by iron deposition. This was confirmed by the level of iron saturation in his blood, the positive genetic test for *C282Y*, and an MRI that showed the iron deposition in his pituitary, pancreas, and liver. Iron in his knuckle joints caused (for reasons unclear) the deposition of calcium dipyrophosphate crystals, which caused damage to the cartilage of the third and fourth knuckle joints. Last, the excess iron stores made him more susceptible to infection with organisms that are supercharged in an environment of iron. *Vibrio*, a bacterium often found in shellfish, is an organism that thrives in an iron-rich environment, like blood.

The good news is that Mr. B. was diagnosed before too much permanent damage had set in. He is on a course of treatment in which he has a unit or two of blood removed each month, which slowly but surely removes some of the iron from his body. More good news is that his children have been tested, and none of them has two of the recessive *C282Y* alleles of the *HFE* gene. None of his children will have the disease, but they could pass it on to their offspring.

I stumbled on the diagnosis, not putting together the multiple clues. I should have done better, as the prevalence of the illness is around five people in a thousand, making it one of the most common genetic diseases in America. My ability to synthesize his multiple, but related, complaints was inadequate. But, fortunately, Mr. B. returned to my office and followed up when he did not get any better with time. His compliance, more than my competence, saved him from a worse outcome.

AN OCTOPUS POT, VOODOO, AND CHANG AND ENG

THEY WERE THE BOSTON EQUIVALENT OF THE residents of Downton Abbey. He was a Harvard grad who, after a career in the U.S. Marines, went to Harvard Divinity School and became pastor of a successful and charitable suburban Episcopal congregation. She was a Radcliffe graduate who raised the children and supported her husband's careers while volunteering and doing good works. They were decent, honest people who saw the good in others and loved each other dearly. And they looked the part: He was stocky and white haired with a patrician accent. She was petite, her white hair was often in braids, she wore simple but elegant clothing, and she had an upper-class Oyster Bay, New York, accent. They were a lovely couple, and their visits to my office were often like an oasis in the desert on a busy day. I always looked forward to seeing them.

Both were quite healthy for people in their early seventies. He had mild hypertension, was moderately overweight, and was wrestling with his recent retirement from his congregation, trying to determine his role in the community and what he wanted to do now that he was no longer on call for his parish 24/7. She was also basically healthy, with mild elevated cholesterol in spite of her slender figure, and with hypertension that was somewhat harder to control. Her main issue was how to deal with her restless husband, now that he was home every day for lunch.

There was some money in the family tree, and they had a lovely summer home on the Maine coast, where they planned to spend even more time. With six grandchildren in their large home, it sounded chaotic to me, but it made them very happy. On their most recent visit, they had asked me if I could find them a physician up there, as they were considering making the summer house their year-round residence.

My impression of their happy and idyllic life was interrupted late one summer afternoon by a call from Susan, the wife. She was very upset. Her husband, Cliff, had come in from a walk along the coast complaining of substernal chest pain. She insisted that I talk with him. Reluctantly, I agreed, but I urged her to call 911 with her cell phone while I was talking to Cliff. Indeed, his symptoms were ominous. He had been walking and throughout the walk felt some chest pain. But, being an ex-marine, he thought he would tough it out, and he kept going. The pain was worse when he negotiated a small hill near his home, and by the time he got into the house, he was sweaty and the pain was unrelenting. I told him I was concerned that this was a heart attack in the making and that he should lie down, quickly take an aspirin, and await the EMTs. He put Susan back on the phone, and I told her the same thing, and again I told her to immediately call 911. I asked her to please update me later in the day, and I gave her my cell phone number.

Around two hours later, Susan called me from Maine Medical Center in Portland wondering if it was safe for Cliff to undergo cardiac catheterization there. I assured her it was, that time was of the essence, and they should proceed with all that the local doctors recommended. I then called the attending cardiologist and gave him the minimal background of Cliff's medical history and his medication list. Some four hours later, I was gratified to hear from the attending cardiologist that, although Cliff had had a heart attack, it was a small one. The thrombosis had been lysed (a clot-dissolving drug had been delivered directly to the thrombosis, which is a blood clot), that two stents had been deployed (these are tiny mesh tubes that are inserted to keep narrowed arteries open), and that he was doing well in the coronary care unit.

When I spoke with Susan, I tried to reassure her as best I could that Cliff's problem was caught in time, that he would do well, and that their life would resume much the same way as before, with the exception of more graded exercise for Cliff and for a host of routine medications he would be given after stenting and a heart attack. But Susan was truly inconsolable. She was weeping, blaming herself for letting Cliff go on his walk alone and for his poor dietary choices. As I listened, I realized she was greatly overwhelmed by this event. I asked her who was with her at the house. Two of the grandchildren were college age, and I thought they would be mature enough to help console and look after her. She was so bereft, and it was so unusual for me to hear her so emotionally upset, that I was not sure she could care for herself.

So I was not entirely surprised when, seven hours later, I received a call from one of the grandchildren who said that Susan wasn't doing well and asked if I would please speak with her. Susan was weeping, and it became clear that she had felt increasing anguish over the last few hours. But in the last thirty minutes, she had developed chest pain. In talking with her, it sounded like classic cardiac pain. She had substernal heaviness, radiation of the pain to her left neck and upper lateral left arm, she felt she might pass out, and she was sweating profusely. I got the granddaughter back on the phone, told her to give Susan an aspirin and, once again, to call the EMTs.

Sometimes women do not have the classic symptoms of heart attacks, but there was no doubt in my mind that Susan was having a cardiac event. Of course, it wasn't lost on me that these symptoms were exactly the same as her husband's. Again, I awaited a call from an attending cardiologist in Portland, and I got it about five hours later. But I was surprised at the diagnosis.

Susan had Takotsubo cardiomyopathy. For once, the medical gobbledygook presented a useful image. Susan did not have the usual cause of a heart attack—namely, a clot in one of the coronary arteries. Rather, she had a sudden malfunction of the muscles of her heart (cardiomyopathy) that took the classic shape of a Japanese octopus trap, or pot, which is what the *takotsubo* means. An octopus pot allows the octopus to enter the trap, but does not let it escape. An angiogram is a radiologic dynamic picture of the heart that is obtained during cardiac catherterization, and an angiogram of a person with Susan's condition shows a left ventricle in the shape of an octopus pot. Although it was first described in 1990 in Japan, this condition is relatively common throughout the world. It tends to affect women in the vast majority of cases, usually women over sixty-five. The good news with Takotsubo cardiomyopathy is that, although the electrocardiogram and cardiac enzyme tests indicate a heart attack, the prognosis is very good. Most patients recover fully, with minimal to no permanent cardiac damage. The most frequent precipitate of the attack, far and away, is stress.

What I never realized in my interactions with Susan, who always seemed so well put together, is that she internalized a great deal and carried a lot of guilt—without any substantive basis for the guilt—and was frequently highly stressed. Unusually, in my relationship with her and her husband, he was more vocal about his emotional state. I was happy to hear of Susan's diagnosis, as I knew the prognosis was good, and it reminded me of the role stress can play in overtly medical, and not psychological, conditions.

This was first impressed upon me when I was an intern. Dr. Bernard Lown had brought in a Haitian voodoo practitioner to sit outside the early coronary care unit in our hospital. One of the patients in the unit was a Haitian man who thought a spell had been cast upon him. Dr. Lown had brought the voodoo doctor to the man's bedside, just to talk with him, and the man went into cardiac arrest. Fortunately, Dr. Lown had co-invented the cardiac defibrillator, and the combination of applying electricity to the patient's heart and removing the voodoo doctor from the room brought the patient back.

I later read about the case of the original Siamese twins, Chang and Eng. They were joined at the sternum and shared a liver, but they did not have a common circulatory system. They eventually emigrated from Siam to North Carolina and married two women. All lived in the same house and shared the same bed until the wives had a falling out, after which they established two households. Chang, later in life, had a stroke, and started to drink heavily. Some four years later, he died, and when Eng awoke, attached to the now dead Chang, he too suddenly died. Autopsy never revealed Eng's cause of death, but my colleague Dr. Martin Samuels, chair of neurology at Brigham and Women's Hospital, makes the compelling case that it was sudden death due to shock or stress. Whether it was due to what we now call Takotsubo cardiomyopathy or some other stress-related cause of death, no one can tell at this date.

Last, Susan and Cliff's situation reminded me of the myth I read long ago when studying Latin. Philemon and Baucis were two kind peasants who were very much in love. Jupiter paid them a visit, and in spite of having little, they fed him freely. After he identified himself, Jupiter asked them what they wished for, and they said that they wished, eventually, to die together. After they served the gods on Mount Olympus for some time, when their time had come, Jupiter simultaneously turned one into an oak tree and the other into a linden tree whose branches became entwined forever.

Driving home the day Susan and Cliff both had heart attacks, thinking of octopus pots, enduring love, voodoo, and Siamese twins, I marveled at the window into life the practice of medicine provides.

ABOUT THE AUTHOR

STUART MUSHLIN, MD, FACP, RECEIVED HIS
medical degree from Weill Cornell Medical College of Cornell University. He worked as an internist and primary care doctor for many years at Brigham and Women's Hospital, a teaching hospital of Harvard Medical School. He is the Master Clinician in Internal Medicine and Primary Care at Brigham and Women's Hospital. He is an Assistant Professor of Medicine at Harvard Medical School and is an Assistant Director of the Residency Program in Internal Medicine at Brigham and Women's Hospital. His numerous publications include a textbook, *Decision Making in Medicine: An Algorithmic Approach*, and research published in the *New England Journal of Medicine*, *Academic Medicine*, *Harvard Health Letter*, and *Arthritis and Rheumatology*. He resides in Brookline, Massachusetts, with his wife.